AF572301

ENVIRONANOTECHNOLOGY

ENVIRONANOTECHNOLOGY

By
Nisha Gupta
P K Gaur

2012

SBS Publishers & Distributors Pvt. Ltd.
New Delhi

ISBN 13 : 9789380090542

First Published in 2012

Published by:

SBS PUBLISHERS & DISTRIBUTORS PVT. LTD.
2/9, Ground Floor, Ansari Road, Darya Ganj,
New Delhi - 110002,
INDIA
Tel: 0091.11.23289119 / 41563911 / 32945311
Email: mail@sbspublishers.com
www.sbspublishers.com

Printed in India by Chaman Enterprises, New Delhi.

Preface

Understanding and utilizing the interactions between environment and nanoscale materials is a new way to resolve the increasingly challenging environmental issues we are facing and will continue to face. Environanotechnology is the nanoscale technology developed for monitoring the quality of the environment, treating water and wastewater, as well as controlling air pollutants. Therefore, the applications of nanotechnology in environmental engineering have been of great interest to many fields and consequently a fair amount of research on the use of nanoscale materials for dealing with environmental issues has been conducted.

Author

Content

Preface ix

1. **Nanoscience in Environment** 1
   - Applications of Nanotechnology
   - Sources of Nanomaterials in the Environment
   - Properties of Nanomaterials
   - Nanomaterial Structure-Toxicity Relationship
   - Potential for Human Exposure
2. **Characterization of Nanomaterials** 48
   - Physicochemical Characterization
   - Characterization Prioritization
   - Testing Methods
   - Conclusion
3. **Natural Colloids and Nanoparticles in Aquatic and Terrestrial Environments** 98
   - Major Types of Environmental Colloids
   - Intrinsic Properties of Environmental Colloidal Particles
   - Interaction Forces Between Soil
4. **Nonorganic Matter** 114
   - Soil Organic Matter
   - The Soil Food
   - Decomposition Process
   - Compounds and Function of Humus

- Natural Factors
- Soil Moisture and Water Saturation
- Soil Texture
- Topography
- Practices that Influence the Amount of Organic Matter
- Decrease in Biomass Production
- Decrease in Organic Matter Supply
- Increased Decomposition Rates
- Practices that Increase Soil Organic Matter
- Increased Biomass Production
- Increased Organic Matter Supply
- Decreased Decomposition Rates
- Effect of Soil Organic Matteron Soil Properties
- Inefficient Use of Rainwater
- Increased Soil Moisture
- Reduced Soil Erosion and Improved Water Quality

5. Atmospheric Nanoparticles 177

- Introduction
- Atmospheric Nanoparticle Formation and Growth: Models and Observations
- Sources of Atmospheric Nanoparticles
- Chemical Composition of Atmospheric Nanoparticles
- Fate and Behaviour of Atmospheric Nanoparticles
- Atmospheric Concentrations

6. Analysis and Characterization of Nanoparticles in Aquatic Environments 214

- Introduction
- Surface Chemical Analysis
- Electron Spectroscopies
- Ion-Based Surface Analysis Methods
- Scanning Probe Microscopies
- Characterization of Carbon Nanotubes (CNTs)

- Considerations
- Sample Mounting Issues
- Analysis Considerations for Specific Methods

7. **Ecotoxicology of Manufactured Nanoparticles** 248
    - Introduction
    - Physico-Chemical Transformation of Nanoparticles
    - Mechanisms of Nanoparticle Toxicity in the Environment
    - Methods
    - Development of Valid/Realistic Toxicity Testing Protocols

8. **Exposure to Nanoparticles** 272
    - Introduction
    - Physical Characteristics and Properties of
    - Nanoparticles
    - Nanoparticle Exposure
    - Control of Exposure

Index 298

1

Nanoscience in Environment

APPLICATIONS OF NANOTECHNOLOGY

Technology-as-we-know-it is a product of industry, of manufacturing and chemical engineering. Industry-as-we-know-it takes things from nature—ore from mountains, trees from forests—and coerces them into forms that someone considers useful. Trees become lumber, then houses. Mountains become rubble, then molten iron, then steel, then cars. Sand becomes a purified gas, then silicon, then chips. And so it goes. Each process is crude, based on cutting, stirring, baking, spraying, etching, grinding.

TREES, THOUGH, ARE NOT CRUDE

To make wood and leaves, they neither cut, grind, stir, bake, spray, etch, nor grind. Instead, they gather solar energy using molecular electronic devices, the photosynthetic reaction centers of chloroplasts. They use that energy to drive *molecular machines*—active devices with moving parts of precise, molecular structure—which process carbon dioxide and water into oxygen and molecular building blocks. They use other molecular machines to join these molecular building blocks to form roots, trunks, branches, twigs, solar collectors, and more molecular machinery. Every tree makes leaves, and each leaf is more sophisticated than a spacecraft, more finely patterned than the latest chip from Silicon Valley. They do all this without noise, heat, toxic fumes, or human labour, and they consume pollutants as they go. Viewed this way, trees are high technology. Chips and rockets aren't.

Trees give a hint of what molecular nanotechnology will be like, but nanotechnology won't be biotechnology because it won't rely on altering life. Biotechnology is a further stage in the domestication of living things. Like selective breeding, it reshapes the genetic heritage of a species to produce varieties more useful to people. Unlike selective breeding, it inserts new genes. Like biotechnology—or ordinary trees—molecular nanotechnology will use molecular machinery, but unlike biotechnology, it will not rely on genetic meddling. It will be not an extension of biotechnology, but an alternative or a replacement.

Molecular nanotechnology could have been conceived and analysed—though not built—based on scientific knowledge available forty years ago. Even today, as development accelerates, understanding grows slowly because molecular nanotechnology merges fields that have been strangers: the molecular sciences, working at the threshold of the quantum realm, and mechanical engineering, still mired in the grease and crudity of conventional technology. Nanotechnology will be a technology of new molecular machines, of gears and shafts and bearings that move and work with parts shaped in accord with the wave equations at the foundations of natural law. Mechanical engineers don't design molecules. Molecular scientists seldom design machines. Yet a new field will grow—is growing today—in the gap between. That field will replace both chemistry as we know it and mechanical engineering as we know it. And what is manufacturing today, or modern technology itself, but a patchwork of crude chemistry and crude machines?

Picture an automated factory, full of conveyor belts, computers, rollers, stampers, and swinging robot arms. Now imagine something like that factory, but a million times smaller and working a million times faster, with parts and workpieces of molecular size. In this factory, a "pollutant" would be a loose molecule, like a ricocheting bolt or washer, and loose molecules aren't tolerated. In many ways, the factory is utterly unlike a living cell: not fluid, flexible, adaptable, and fertile, but rigid, preprogrammed and specialized. And yet for all of that, this microscopic molecular factory emulates life in its clean, precise molecular construction.

Advanced molecular manufacturing will be able to make almost anything. Unlike crude mechanical and chemical technologies, molecular manufacturing will work from the bottom up, assembling intricate products from the molecular building blocks that underlie everything in the physical world.

Nanotechnology will bring new capabilities, giving us new ways to make things, heal our bodies, and care for the environment. It will also bring unwelcome advances in weaponry and give us yet more ways to foul up the world on an enormous scale. It won't automatically solve our problems: even powerful technologies merely give us more power. As usual, we have a lot of work ahead of us and a lot of hard decisions to make if we hope to harness new developments to good ends. The main reason to pay attention to nanotechnology now, before it exists, is to get a head start on understanding it and what to do about it.

CONSEQUENCES

The United States has become famous for its obsession with the next year's elections and the next quarter's profits, and the future be damned. Nonetheless, we are writing for normal human beings who feel that the future matters–ten, twenty, perhaps even thirty years from now—for people who care enough to try to shift the odds for the better. Making wise choices with an eye to the future requires a realistic picture of what the future can hold. What if most pictures of the future today are based on the wrong assumptions?

Here are a few of today's common assumptions, some so familiar that they are seldom stated:

- Industrial development is the only alternative to poverty.

- Many people must work in factories.
- Greater wealth means greater resource consumption.
- Logging, mining, and fossil-fuel burning must continue.
- Manufacturing means polluting.
- Third World development would doom the environment.

Some Further Common Assumptions

- The twenty-first century will basically bring more of the same.
- Today's economic trends will define tomorrow's problems.
- Spaceflight will never be affordable for most people.
- Forests will never grow beyond Earth.
- More advanced medicine will always be more expensive.
- Even highly advanced medicine won't be able to keep people healthy.
- Solar energy will never become really inexpensive.
- Toxic wastes will never be gathered and eliminated.
- Developed land will never be returned to wilderness.
- There will never be weapons worse than nuclear missiles.
- Pollution and resource depletion will eventually bring war or collapse.

These commonplace assumptions paint a future full of terrible dilemmas, and the notion that a technological change will let us escape from them smacks of the idea that some technological fix can save the industrial system. The prospect, though, is quite different: The industrial system won't be fixed, it will be junked and recycled. The prospect isn't more industrial wealth ripped from the flesh of the Earth, but green wealth unfolding from processes as clean as a growing tree. Today, our industrial technologies force us to choose better quality *or* lower cost*or* greater safety *or* a cleaner environment. Molecular manufacturing, however, can be used to improve quality *and* lower costs *and* increase safety *and* clean the environment. The coming revolutions in technology will transcend many of the old, familiar dilemmas. And yes, they will bring fresh, equally terrible dilemmas.

Molecular nanotechnology will bring thorough and inexpensive control of the structure of matter. We need to understand molecular nanotechnology in order to understand the future capabilities of the human race. This will help us see the challenges ahead, and help us plan how best to conserve values, traditions, and ecosystems through effective policies and institutions. Likewise, it can help us see what today's events mean, including business opportunities and possibilities for action. We need a vision of where technology is leading because technology is a part of what human beings are, and will affect what we and our societies can become.

The consequences of the coming revolutions will depend on human actions. As always, new abilities will create new possibilities both for good and for ill. We will discuss both, focusing on how political and economic pressures can best be harnessed to achieve good ends. Our answers will not be satisfactory, but they are at least a beginning.

TRENDS

Technology has been moving towards greater control of the structure of matter for millennia. For decades, microtechnology has been building ever-smaller devices, working towards the molecular size scale from the top down. For a century or more, chemistry has been building ever-larger molecules, working up towards molecules large enough to serve as machines. The research is global, and the competition is heating up.

Naturally occurring molecular machines exist already. Researchers are learning to design new ones. The trend is clear, and it will accelerate because better molecular machines can help build even better molecular machines. By the standards of daily life, the development of molecular nanotechnology will be gradual, spanning years or decades, yet by the ponderous standards of human history it will happen in an eyeblink. In retrospect, the wholesale replacement of twentieth-century technologies will surely be seen as a technological revolution, as a process encompassing a great breakthrough.

Today, we live in the end of the pre-breakthrough era, with pre-breakthrough technologies, hopes, fears, and preoccupations that often seem permanent, as did the Cold War. Yet it seems that the breakthrough era is not a matter for some future generation, but for our own. These developments are taking shape right now, and it would be rash to assume that their consequences will be many years delayed.

To get a sense of the consequences, though, requires a picture of what nanotechnology can do. This can be hard to grasp because past advanced technologies–microwave tubes, lasers, superconductors, satellites, robots, and the like–have come trickling out of factories, at first with high price tags and narrow applications. Molecular manufacturing, though, will be more like computers: a flexible technology with a huge range of applications. And molecular manufacturing won't come trickling out of conventional factories as computers did: it will *replace* factories and replace or upgrade their products. This is something new and basic, not just another twentieth-century gadget. It will arise out of twentieth-century trends in science, but it will break the trend-lines in technology, economics, and environmental affairs.

Calculators were once thousand-dollar desktop clunkers, but microelectronics made them fast and efficient, sized to a child's pocket and priced to a child's budget. Now imagine a revolution of similar magnitude, but applied to everything else.

More Consequences: Scenes from a Post-Breakthrough World

What nanotechnology will mean for human life is beyond our predicting, but a good way to understand what it *could* mean is to paint scenarios. A good scenario brings together different aspects of the world (technologies, environments, human concerns) into a coherent whole.

The following scenarios can't represent what will happen, because no one knows. They can, however, show how post-breakthrough capabilities could mesh with human life and Earth's environment. The results will likely seem quaintly conservative from a future perspective, however much they seem like science fiction today.

SOLAR ENERGY

In Fairbanks, Alaska, Linda Hoover yawns and flips a switch on a dark winter morning. The light comes on, powered by stored solar electricity. The Alaska oil pipeline shut down years ago, and tanker traffic is gone for good.

Nanotechnology can make solar cells efficient, as cheap as newspaper, and as tough as asphalt–tough enough to use for resurfacing roads, collecting energy without displacing any more grass and trees. Together with efficient, inexpensive storage cells, this will yield low-cost power (but no, not "too cheap to meter").

SCENARIO: MEDICINE THAT CURES

Sue Miller of Lincoln, Nebraska, has been a bit hoarse for weeks, and just came down with a horrid head cold. For the past six months, she's been seeing ads for At Last!: the Cure for the Common Cold, so she spends her five dollars and takes the nose-spray and throat-spray doses. Within three hours, 99 per cent of the viruses in her nose and throat are gone, and the rest are on the run. Within six hours, the medical mechanisms have become inactive, like a pinch of inhaled but biodegradable dust, soon cleared from the body. She feels much better and won't infect her friends at dinner.

The human immune system is an intricate molecular mechanism, patrolling the body for viruses and other invaders, recognizing them by their foreign molecular coats. The immune system, though, is slow to recognize something new. For her five dollars, Sue bought 10 billion molecular mechanisms primed to recognize not just the viruses she had already encountered, but each of the five hundred most common viruses that cause colds, influenza, and the like. Weeks have passed, but the hoarseness Sue had before her cold still hasn't gone away; it gets worse. She ignores it through a long vacation, but once she's back and caught up, Sue finally goes to see her doctor. He looks down her throat and says, "Hmmm." He asks her to inhale an aerosol, cough, spit in a cup, and go read a magazine. The diagnosis pops up on a screen five minutes after he pours the sample into his cell analyser. Despite his knowledge, his training and tools, he feels chilled to read the diagnosis: a malignant cancer of the throat, the same disease that has cropped up all too often in his own mother's family.

He touches the "Proceed" button. In twenty minutes, he looks at the screen to check progress. Yes, Sue's cancerous cells are all of one basic kind, displaying one of the 16,314 known molecular markers for malignancy. They can be recognized, and since they can be recognized, they can be destroyed by standard molecular machines primed to react to those markers. The doctor instructs the cell analyser to prime some "immune machines" to go after her cancer cells. He tests them on cells from the sample, watches, and sees that they work as expected, so he has the analyser prime up some more.

CLEANSING THE SOIL

California Scout Troop 9731 has hiked for six days, deep in the second-wilderness forests of the Pacific Northwest. "I bet we're the first people ever to walk here," says one of the youngest scouts. "Well, maybe you're right about *walking*," says Scoutmaster Jackson, "but look up ahead–what do you see, scouts?" Twenty paces ahead runs a strip of younger trees, stretching left and right until it vanishes among the trunks of the surrounding forest.

"Hey, guys! Another old logging road!" shouts an older scout. Several scouts pull probes from their pockets and fit them to the ends of their walking sticks. Jackson smiles: It's been ten years since a California troop found anything this way, but the kids keep trying. The scouts fan out, angling their path along the scar of the old road, poking at the ground and watching the readouts on the stick handles. Suddenly, unexpectedly, comes a call: "I've got a signal! Wow–I've got PCBs!"

In a moment, grinning scouts are mapping and tracing the spill. Decades ago, a truck with a leaking load of chemical waste snuck down the old logging road, leaving a thin toxic trail. That trail leads them to a deep ravine, some rusted drums, and a nice wide patch of invisible filth. The excitement is electrifying.

Setting aside their maps and orienteering practice, they unseal a satellite locator to log the exact latitude and longitude of the site, then send a message that registers their cleanup claim on the ravine. The survey done, they head off again, eagerly planning a return trip to earn the now-rare Toxic Waste Cleanup Merit Badge. Today, tree farms are replacing wilderness. Tomorrow, the slow return to wilderness may begin, when nature need no longer be seen as a storehouse of natural resources to be plundered.

SUPERCOMPUTERS

At the University of Michigan, Joel Gregory grabs a molecular rod with both hands and twists. It feels a bit weak, and a ripple of red reveals too much stress in a strained molecular bond halfway down its length. He adds two atoms and twists the rod again: all greens and blues, much better. Joel plugs the rod into the mechanical arm he's designing, turns up the temperature, and sets the whole thing in motion. A million atoms dance in thermal vibration, gears spin, and the arm swings to and fro in programmed motion. Joel strips off the computer display goggles and gloves and blinks at the real world. It's time for a sandwich and a cup of coffee. He grabs the computer itself, stuffs it into his pocket, and heads for the student center.

Researchers already use computers to build models of molecules, and "virtual reality systems" have begun to appear, enabling a user to walk around the image of a molecule and "touch" it, using computer-controlled gloves and goggles. We can't build a supercomputer able to model a million-atom machine yet–much less build a pocket supercomputer–but computers keep shrinking in size and cost. With nanotechnology to make molecular parts, a computer like Joel's will become easy to build.

GLOBAL WEALTH

Behind a village school in the forest a stone's throw from the Congo River, a desktop computer with a thousand times the power of an early 1990s supercomputer lies half-buried in a recycling bin. Indoors, Joseph Adoula and his friends have finished their day's studies; now they are playing together in a vivid game universe using personal computers each a million times more powerful than the clunker in the trash. They stay late in air-conditioned comfort.

Trees use air, soil, and sunlight to make wood, and wood is cheap enough to burn. Nanotechnology can do likewise, making products as cheap as wood–even products like supercomputers, air conditioners, and solar cells to power them. The resulting economics may even keep tropical forests from being burned.

CLEANSING THE AIR

In Earth's atmosphere, the twentieth-century rise in carbon-dioxide levels has halted and reversed. Fossil fuels are obsolete, so pollution rates have lessened. Efficient agriculture has freed fertile land for reforestation, so growing trees are cleansing the atmosphere. Surplus solar power from the world's repaved roads is being used to break down excess carbon dioxide at a rate of 5 billion tons per year. Climates are returning to normal, the seas are receding to their historical shores, and ecosystems are beginning the slow process of recovery. In another twenty years, the atmosphere will be back to the pre-industrial composition it had in the year 1800.

TRANSPORTATION OUTWARD

Jim Salin's afternoon flight from Dulles International is on the ground, late for departure. Impatiently, Jim checks the time: any later, and he'll miss his connecting flight.

At last, the glassy-surfaced craft rolls down the runway. With gliderlike wings, it lifts its fat body and climbs steeply towards the east. A few pages into his novel, Jim is interrupted by a second recitation of safety instructions and the captain's announcement that they'll try to make up for lost time. Jim settles back in his seat as the main engines kick in, the wings retract, the acceleration builds, and the sky darkens to black. Like the highest-performance rockets of the 1980s, Jim's liner produces an exhaust of pure water vapour. Spaceflight has become clean, safe, and routine. And every year, more people go up than come down.

The cost of spaceflight is mostly the cost of high-performance, reliable hardware. Molecular manufacturing will make aerospace structures from nearly flawless, superstrong materials at low cost. Add inexpensive fuel, and space will become more accessible than the other side of the ocean is today.

RESTORING SPECIES

Restoration Day Ceremonies are always moving events. For some reason, the old people always cry, even though they say they're happy. *Crying,* Tracy Stiegler thinks, *doesn't make any sense.* She looks again through the camouflage screen over the sandy Triangle Keys beach, gazing across the Caribbean towards the Yucatán Peninsula. *Soon this will be theirs again, and that's all to the good.*

Tracy and the other scientists from BioArchive have positions of honour in today's Restoration Day Ceremony. Since the mid-twentieth century there had been no living Caribbean monk seals, only grisly relics of the years of their slaughter: seal furs and dry museum specimens. Tracy's team struggled for years, gathering these relics and studying them with molecular instruments. It had been known for decades—since the 1980s—that genes are tough enough to survive in dried skin, bone, horn, and eggshell. Tracy's team had collected genes and rebuilt cells.

They worked for years, and gave thanks to the strict protection—late, but good enough—that saved one related species. At last, a Hawaiian monk seal had given birth to a genetically-pure Caribbean monk seal, twin to a seal long dead. And now there were five hundred, some young, some middle-aged, with decent genetic diversity and five years' experience living in the confines of a coastal ecological station.

Today, with raucous voices, they are moving out into the world to reclaim their ecological niche. As Tracy watches, she thinks of the voices that will never be heard again: of the species, known and unknown, that left not a even a bloody scrap to be cherished and restored. Thousands (millions?) of species had simply been brushed into extinction as habitats were destroyed by farming and logging. People knew–for *years* they had known–that freezing or drying would save genes. And they knew of the ecological destruction, and they knew they weren't stopping it. And the ignorant bastards didn't even keep samples.

People will surely push biomedical applications of nanotechnology far and fast for human health-care. With a bit more pushing, this technology base will be good enough to restore some species now thought lost forever, to repair some of the damage human beings have done to the web of life. It would be better to preserve ecosystems and species intact, but restoration, even of a few species, will be far better than nothing. Some samples from endangered species are being kept today, but not enough, and mostly for the wrong reasons.

UNSTABLE ARMS RACE

Disputes over technology development and trade had soured relationships between Singapore and the Japan-United States alliance. Diplomatic enquiries regarding peculiar seismic and sonar readings in the South China Sea had just begun when they suddenly became irrelevant: an estimated one billion tons of unfamiliar, highly-automated military hardware appeared in coastal waters around the world. Accusations began to fly between Congress and PeaceWatch personnel.

Molecules matter because matter is made of molecules, and everything from air to flesh to spacecraft is made of matter. When we learn how to arrange molecules in new ways, we can make new things, and make old things in new ways. Perhaps this is why Japan's MITI has identified "control technologies for the precision arrangement of molecules" as a basic industrial technology for the twenty-first century. Molecular nanotechnology will give thorough control of matter on a large scale at low cost, shattering a whole set of technological and economic barriers more or less at one stroke.

A molecule is an object consisting of a collection of atoms held together by strong bonds (one-atom molecules are a special case). "Molecule" usually refers to an object with a number of atoms small enough to be counted (a few to a few thousand), but strictly speaking a truck tire (for instance) is mostly one big molecule, containing something like 1,000,000,000,000,000,000, 000,000,000 atoms. Counting this many atoms aloud would take about 10,000,000,000 *billion* years.

Scientists and engineers still have no direct, convenient way to control molecules, basically because human hands are about 10 million times too large. Today, chemists and materials scientists make molecular structures indirectly, by mixing, heating, and the like. The idea of nanotechnology begins with the idea of a *molecular* assembler, a device resembling an industrial robot arm but built on a microscopic scale. A general-purpose molecular assembler will be a jointed mechanism built from rigid molecular parts, driven by motors, controlled by computers, and able to grasp and apply molecular-scale tools. Molecular assemblers can be used to build other molecular machines–they can even build more molecular assemblers. Assemblers and other machines in molecular manufacturing systems will be able to make almost anything, if given the right raw materials. In effect, molecular assemblers will provide the microscopic "hands" that we lack today.

Nanotechnology will give better control of molecular building blocks, of how they move and go together to form more complex objects. Molecular manufacturing will make things by building from the bottom up, starting with the smallest possible building blocks. The *nano* in nanotechnology comes from *nanos*, the Greek word for dwarf. In science, the prefix nano-means one-billionth of something, as in nanometer and nanosecond, which are typical units of size and time in the world of molecular manufacturing. When you see it tacked onto the name of an object, it means that the object is made by patterning matter with molecular control:nanomachine, nanomotor, minicomputer. These are the smallest, most precise devices that make sense based on today's science.

(Be cautious of other usages, though—some researchers have begun to use the *nano-* prefix to refer to other small-scale technologies in the laboratory today. In this *nanotechnology* means the *precise, molecular* nanotechnology of the future. British usage also applies the term to the small-scale and high precision technologies of today—even to precision grinding and measurement. The latter are useful, but hardly revolutionary.)

Digital electronics brought an information-processing revolution by handling information quickly and controllably in perfect, discrete pieces: bits and bytes. Likewise, nanotechnology will bring a matter-processing revolution by handling matter quickly and controllably in perfect, discrete pieces: atoms and molecules. The digital revolution has centered on a device able to make any desired pattern of bits: the programmable computer. Likewise, the nanotechnological revolution will center on a device able to make (almost) any desired pattern of atoms: the programmable assembler. The technologies that plague us today suffer from the messiness and wear of an old phonograph record. Nanotechnology, in contrast, will bring the crisp, digital perfection of a compact disc.

NANOFABRICATION

The next room displays more technology. Here in the workroom where SuperChips were first made, early nanotech manufacturing is spread out on display. The whole setup is surprisingly quiet and ordinary. Back in the 1980s and 1990s, chip plants had carefully controlled clean rooms with gowns and masks on workers and visitors, special workstations, and carefully crafted air flows to keep dust away from products. This room has none of that. It's even a little grubby. In the middle of a big square table are a half-dozen steel tanks, about the size and shape of old-fashioned milk cans. Each can has a different label identifying its contents: memory blocks, data-transmission blocks, interface blocks. These are the parts needed for building up the chip. Clear plastic tubes, carrying clear and tea-coloured liquids, emerge from the mouths of the milk cans and drape across the table. The tubes end in fist-sized boxes mounted above shallow dishes sitting in a ring around the cans. As the different liquids drip into each dish, a beater like a kitchen mixer swirls the liquid.

A Nanofab "engineer," dressed in period clothing complete with name badge, is setting up a dish to begin building a new chip. "This," he says, holding up a blank with a pair of tweezers, "is a silicon chip like the ones made with pre-breakthrough technology. Companies here in this valley made chips like these by melting silicon, freezing it into lumps, sawing the lumps into slices, polishing the slices, and then going through a long series of chemical and photographic steps. When they were done, they had a pattern of lines and blobs of different materials on the surface. Even the smallest of these blobs contained *billions* of atoms, and it took several blobs working together to store a single bit of information. A chip this size, the size of your fingernail, could store only a fraction of a billion bits. Here at Nanofab, we used bare silicon chips as a base for building up nanomemory. The picture on the wall here shows the surface of a blank chip: no transistors, no memory circuits, just fine wires to connect up with the nanomemory we built on top. The nanomemory, even in the early days, stored*thousands* of billions of bits. And we made them like this, but a thousand at a time–" He places the chip in the dish, presses a button, and the dish begins to fill with liquid.

The chips in the dishes all look pretty much the same except for colour. The new chip looks like dull metal. The only difference you can see in the older chips, further along in the process, is a smooth rectangular patch covered by a film of darker material. An animated flowchart on the wall shows how layer upon layer of nanomemory building blocks are grabbed from solution and laid down on the surface to make that film.

The total amount of energy needed here is trivial, because the amount of product is trivial: at the end of the process, the total thickness of nanomemory structure–the memory store for a Pocket Library–amounts to one-tenth the thickness of a sheet of paper, spread over an area smaller than a postage stamp.

MOLECULAR ASSEMBLY

The animated flowchart showed nanomemory building blocks as big things containing about a hundred thousand atoms apiece (it takes a moment to remember that this is still submicroscopic). The build process in the dishes stacked these blocks to make the memory film on the SuperChip, but how were the blocks themselves built? The hard part in this molecular-manufacturing business has got to be at the bottom of the whole process, at the stage where molecules are put together to make large, complex parts.

The Silicon Valley Faire offers simulations of this molecular assembly process, and at no extra charge. From the tourguide, you learn that modern assembly processes are complex; that earlier processes–like those used by Nanofabricators, Inc. –used clever-but-obscure engineering tricks; and that the simplest, earliest concepts were never built. Why not begin at the beginning? A short walk takes you to the Museum of Antique Concepts, the first wing of the Museum of Molecular Manufacturing.

A peek inside the first hall shows several people strolling around wearing loosely fitting jumpsuits with attached goggles and gloves, staring at nothing and playing mime with invisible objects. Oh well, why not join the fools' parade? Stepping through the doorway while wearing the suit is entirely different. The goggles show a normal world outside the door and a molecular world inside. Now you, too, can see and feel the exhibit that fills the hall. It's much like the earlier simulated molecular world: it shares the standard settings for size, strength, and speed. Again, atoms seem 40 million times larger, about the size of your fingertips. This simulation is a bit less thorough than the last was–you can feel simulated objects, but only with your gloved hands. Again, everything seems to be made of quivering masses of fused marbles, each an atom.

Machinery fills the hall. Overall, the sight is reminiscent of an automated factory of the 1980s or 1990s. It seems clear enough what must be going on: Big machines stand beside a conveyor belt loaded with half-finished-looking blocks of some material; the machines must do some sort of work on the blocks. Judging by the conveyor belt, the blocks eventually move from one arm to the next until they turn a corner and enter the next hall.

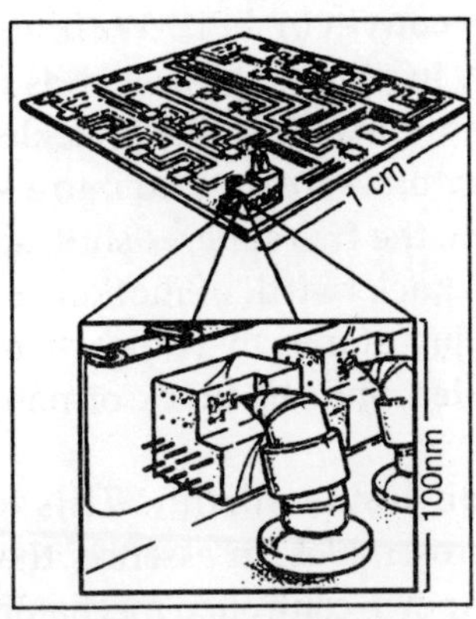

Fig. 1.1 Assembler With Factory on Chip

Since nothing is real, the exhibit can't be damaged, so you walk up to a machine and give it a poke. It seems as solid as the wall of the nanocomputer in the previous tour. Suddenly, you notice something odd: no bombarding air molecules and no droplets of water–in fact, no loose molecules anywhere. Every atom seems to be part of a mechanical system, quivering thermal vibration, but otherwise perfectly controlled. Everything here is like the nanocomputer or like the tough little gear; none of it resembles the loosely coiled protein or the roiling mass of the living cell.

The conveyor belt seems motionless. At regular intervals along the belt are blocks of material under construction: workpieces. The nearest block is about a hundred marble-bumps wide, so it must contain something like 100 × 100 × 100 atoms, a full million. This block looks strangely familiar, with its rods, crank, and the rest. It's a nanocomputer–or rather, a blocklike part of a nanocomputer still under construction.

Standing alongside the pieces of nanocomputer on the conveyor belt, dominating the hall, is a row of huge mechanisms. Their trunks rise from the floor, as thick as old oaks. Even though they bend over, they rear overhead. "Each machine," your tourguide says, "is the arm of a general-purpose molecular assembler. One assembler arm is bent over with its tip pressed to a block on the conveyor belt. Walking closer, you see molecular assembly in action. The arm ends in a fist-sized knob with a few protruding marbles, like knuckles. Right now, two quivering marbles–atoms–are pressed into a small hollow in the block. As you watch, the two spheres shift, snapping into place in the block with a quick twitch of motion: a chemical reaction. The assembler arm just stands there, nearly motionless. The fist has lost two knuckles, and the block of nanocomputer is two atoms larger.

The tourguide holds forth: "This general-purpose assembler concept resembles, in essence, the factory robots of the 1980s. It is a computer-controlled mechanical arm that moves molecular tools according to a series of instructions. Each tool is like a single-shot stapler or rivet gun. It has a handle for the assembler to grab and comes loaded with a little bit of matter–a few atoms–which it attaches to the workpiece by a chemical reaction." This is like the rejoining of the protein chain in the earlier tour.

MOLECULAR PRECISION

The atoms seemed to jump into place easily enough; can they jump out of place just as easily? By now the assembler arm has crept back from the surface, leaving a small gap, so you can reach in and poke at the newly added atoms. Poking and prying do no good: When you push as hard as you can, the atoms don't budge by a visible amount. Strong molecular bonds hold them in place. Your pocket tourguide–which has been applying the power of a thousand 1990s supercomputers to the task of deciding when to speak up–remarks, "Molecular bonds hold things together. In strong, stable materials atoms are either bonded, or they aren't, with no possibilities in between. Assemblers work by making and breaking bonds, so each step either succeeds perfectly or fails completely. In pre-breakthrough manufacturing, parts were always made and put together with small inaccuracies. These could add up to wreck product quality. At the molecular scale, these problems vanish. Since each step is perfectly precise, little errors can't add up. The process either works, or it doesn't."

But what about those definite, complete failures? Fired by scientific curiosity, you walk to the next assembler, grab the tip, and shake it. Almost nothing happens. When you shove as hard as you can, the tip moves by about one-tenth of an atomic diameter, then springs back. "Thermal vibrations can cause mistakes by causing parts to come together and form bonds in the wrong place," the tourguide remarks. "Thermal vibrations make floppy objects bend further than stiff ones, and so these assembler arms were designed to be thick and stubby to make them very stiff. Error rates can be kept to one in a trillion, and so small products can be perfectly regular and perfectly identical. Large products can be*almost* perfect, having just a few atoms out of place." This should mean high reliability. Oddly, most of the things you've been seeing outside have looked pretty ordinary–not slick, shiny, and perfect, but rough and homey. They must have been *manufactured* that way, or made by hand. Slick, shiny things must not impress anyone anymore.

MOLECULAR ROBOTICS

By now, the assembler arm has moved by several atom-widths. Through the translucent sides of the arm you can see that the arm is full of mechanisms: twirling shafts, gears, and large, slowly turning rings that drive the rotation and extension of joints along the trunk. The whole system is a huge, articulated robot arm. The arm is big because the smallest parts are the size of marbles, and the machinery inside that makes it move and bend has many, many parts. Inside, another mechanism is at work: The arm now ends in a hole, and you can see the old, spent molecular tool being retracted through a tube down the middle. Patience, patience. Within a few minutes, a new tool is on its way back up the tube. Eventually, it reaches the end. Shafts twirl, gears turn, and clamps lock the tool in position. Other shafts twirl, and the arm slowly leans up against the workpiece again at a new site. Finally, with a twitch of motion, more atoms jump across, and the block is again just a little bit bigger. The cycle begins again. This huge arm seems amazingly slow, but the standard simulation settings have shifted speeds by a factor of over 400 million. A few minutes of simulation time correspond to less than a millionth of a second of real time, so this stiff, sluggish arm is completing about a million operations per second.

Peering down at the very base of the assembler arm, you can get a glimpse of yet more assembler-arm machinery underneath the floor: Electric motors spin, and a nanocomputer chugs away, rods pumping furiously. All these rods and gears move quickly, sliding and turning many times for every cycle of the ponderous arm. This seems inefficient; the mechanical vibrations must generate a lot of heat, so the electric motors must draw a lot of power. Having a computer control each arm is a lot more awkward now than it was in pre-breakthrough years. Back then, a robot arm was big and expensive and a computer was a cheap chip; now the computer is bigger than the arm. There must be a better way–but then, this is the Museum of Antique Concepts.

Building-Blocks into Buildings

Where do the blocks go, once the assemblers have finished with them? Following the conveyor belt past a dozen arms, you stroll to the end of the hall, turn the corner, and find yourself on a balcony overlooking a vaster hall beyond. Here, just off the conveyor belt, a block sits in a complex fixture. Its parts are moving, and an enormous arm looms over it like a construction crane. After a moment, the tourguide speaks up and confirms your suspicion: "After manufacturing, each block is tested.

Large arms pick up properly made blocks. In this hall, the larger arms assemble almost a thousand blocks of various kinds to make a complete nanocomputer.

The grand hall has its own conveyor belt, bearing a series of partially completed nanocomputers. Arrayed along this grand belt is a row of grand arms, able to swing to and fro, to reach down to lesser conveyor belts, pluck million-atom blocks from testing stations, and plug them into the grand workpieces, the nanocomputers under construction. The belt runs the length of the hall, and at the end, finished nanocomputers turn a corner–to a yet-grander hall beyond?

After gazing at the final-assembly hall for several minutes, you notice that nothing seems to have moved. Mere patience won't do: at the rate the smaller arms moved in the hall behind you, each block must take months to complete, and the grand block-handling arms are taking full advantage of the leisure this provides. Building a computer, start to finish, might take a terribly long time. Perhaps as long as the blink of an eye.

Molecular assemblers build blocks that go to block assemblers. The block assemblers build computers, which go to system assemblers, which build systems, which–at least one path from molecules to large products seems clear enough. If a car were assembled by normal-sized robots from a thousand pieces, each piece having been assembled by smaller robots from a thousand smaller pieces, and so on, down and down, then only ten levels of assembly process would separate cars from molecules. Perhaps, around a few more corners and down a few more ever-larger halls, you would see a post-breakthrough car in the making, with unrecognizable engine parts and comfortable seating being snapped together in a century-long process in a hall so vast that the Pacific Ocean would be a puddle in the corner.

MOLECULAR PROCESSING

Stepping back into that hall, you wonder how the process begins. In every cycle of their sluggish motion, each molecular assembler gets a fresh tool through a tube from somewhere beneath the floor, and that *somewhere* is where the story of molecular precision begins. And so you ask, "Where do the tools come from?", and the tourguide replies, "You might want to take the elevator to your left."

Stepping out of the elevator and into the basement, you see a wide hall full of small conveyor belts and pulleys; a large pipe runs down the middle. A plaque on the wall says, "Mechanochemical processing concept, circa 1990." As usual, all the motions seem rather slow, but in this hall everything that seems designed to move is visibly in motion. The general flow seems to be away from the pipe, through several steps, and then up through the ceiling towards the hall of assemblers above.

After walking over to the pipe, you can see that it is nearly transparent. Inside is a seething chaos of small molecules: the wall of the pipe is the boundary between loose molecules and controlled ones, but the loose molecules are well confined. In this simulation, your fingertips are like small molecules. No matter how hard you push, there's no way to drive your finger through the wall of the pipe. Every few paces along the pipe a fitting juts out, a housing with a mechanically driven rotating thing, exposed to the liquid inside the pipe, but also exposed to a belt running over one of the pulleys, embedded in the housing. It's hard to see exactly what is happening.

The tourguide speaks up, saying, "Pockets on the rotor capture single molecules from the liquid in the pipe. Each rotor pocket has a size and shape that fits just one of the several different kinds of molecule in the liquid, so the process is rather selective. Captured molecules are then pushed into the pockets on the belt that's wrapped over the pulley there, then–"

"Enough," you say. Fine, it singles out molecules and sticks them into this maze of machinery. Presumably, the machines can sort the molecules to make sure the right kinds go to the right places.

The belts loop back and forth carrying big, knobby masses of molecules. Many of the pulleys–rollers?–press two belts together inside a housing with auxiliary rollers. While you are looking at one of these, the tourguide says, "Each knob on a belt is a mechanochemical-processing device. When two knobs on different belts are pressed together in the right way, they are designed to transfer molecular fragments from one to another by means of a mechanically forced chemical reaction. In this way, small molecules are broken down, recombined, and finally joined to molecular tools of the sort used in the assemblers in the hall above. In this device here, the rollers create a pressure equal to the pressure found halfway to the center of the Earth, speeding a reaction that–" "Fine, fine," you say. Chemists in the old days managed to make amazingly complex molecules just by mixing different chemicals together in solution in the right order under the right conditions.

Here, molecules can certainly be brought together in the right order, and the conditions are much better controlled. It stands to reason that this carefully designed maze of pulleys and belts can do a better job of molecule processing than a test tube full of disorganized liquid ever could. From a liquid, through a sorter, into a mill, and out as tools: this seems to be the story of molecule processing. All the belts are loops, so the machinery just goes around and around, carrying and transforming molecular parts.

Beyond Antiques

This system of belts seems terribly simple and efficient, compared to the ponderous arms driven by frantic computers in the hall above. Why stop with making simple tools? You must have muttered this, because the tourguide speaks up again and says, "The Special-Assembler Exhibit shows another early molecular-manufacturing concept that uses the principles of this molecule-processing system to build large, complex objects. If a system is building only a single product, there is no need to have computers and flexible arms move parts around. It is far more efficient to build a machine in which everything just moves on belts at a constant speed, adding small parts to larger ones and then bringing the larger ones together as you saw at the end of the hall above." This does seem like a more sensible way to churn out a lot of identical products, but it sounds like just more of the same. Gears like fused marbles, belts like coarse beadwork, drive shafts, pulleys, machines and more machines. In a few places, marbles snap into new patterns to prepare a tool or make a product. Roll, roll, chug, chug, pop, snap, then roll and chug some more.

The tourguide launches into a list: "Yes–the inner workings of assembler arms, with drive shafts, worm gears, and harmonic drives; the use of Diels-Alder reactions, interfacial free-radial chain reactions, and dative-bond formation to join blocks together in the larger-scale stages of assembly; different kinds of mechanochemical processing for preparing reactive molecular tools; the use of staged-cascade methods in providing feed-molecules of the right kinds with near-perfect reliability; the differences between efficient and inefficient steps in molecular processing; the use of redundancy to ensure reliability in large systems despite sporadic damage; modern methods of building large objects from smaller blocks; modern electronic nanocomputers; modern methods for–"

CRUDE TECHNOLOGY

As the Silicon Valley Faire scenario shows, molecular manufacturing will work much like ordinary manufacturing, but with devices built so small that a single loose molecule of pollutant would be like a brick heaved into a machine tool. John Walker of Autodesk, a leading company in computer-aided design, observes that nanotechnology and today's crude methods are very different: "Technology has never had this kind of precise control; all of our technologies today are bulk technologies. We take a big chunk of stuff and hack away at it until we're left with the object we want, or we assemble parts from components without regard to structure at the molecular level."

Molecular manufacturing will orchestrate atoms into products of symphonic complexity, but modern manufacturing mostly makes loud noises. These figurative noises are sometimes all too literal: A crack in a metal forging grows under stress, a wing fails, and a passenger jet crashes from the sky. A chemical reaction goes out of control, heat and pressure build, and a poisonous blast shakes the countryside. A lifesaving product cannot be made, a heart fails, and a hospital's heart-monitoring machine signals the end with a high-pitched wail.

Today, we make many things from metal, by machining. From the perspective of our standard, simulated molecular world, a typical metal part is a piece of terrain many days' journey across. The metal itself is weak compared to the bonds of the protein chain or other tough nanomechanisms: solid steel is no stronger than your simulated fingers, and the atoms on its surface can be pushed around with your bare hands. Standing on a piece of metal being machined in a lathe, you would see a cutting blade crawl past a few times per year, like a majestic plough the size of a mountain range. Each pass would rip up a strip of the metal landscape, leaving a rugged valley broad enough to hold a town. This is machining from a nanotechnological perspective: a process that hacks crude shapes from intrinsically weak materials.

Today, electronics are made from silicon chips. We have already seen the landscape of a finished chip. During manufacturing, metal features would be built up by a centuries-long drizzle of metal-atom rain, and hollows would be formed by a centuries-long submergence in an acid sea. From the perspective of our simulation, the whole process would resemble geology as much as manufacturing, with the slow layering of sedimentary deposits alternating with ages of erosion. The term *nanotechnology* is sometimes used as a name for small-scale microtechnology, but the difference between molecular manufacturing and this sort of microlandscaping is like the difference between watchmaking and bulldozing.

Today, chemists make molecules by solution chemistry. We have seen what a liquid looks like in our first simulation, with molecules bumping and tumbling and wandering around. Just as assemblers can make chemical reactions occur by bringing molecules together mechanically, so reactions can occur when molecules bump at random through thermal vibration and motion in a liquid. Indeed, much of what we know today about chemical reactions comes from observing this process. Chemists make large molecules by mixing small molecules in a liquid. By choosing the right molecules and conditions, they can get a surprising measure of control over the results: only some pairs of molecules will react, and then only in certain ways.

Doing chemistry this way, though, is like trying to assemble a model car by putting the pieces in a box and shaking. This will only work with cleverly shaped pieces, and it is hard to make anything very complex. Chemists today consider it challenging to make a precise, three-dimensional structure having a hundred atoms, and making one with a thousand atoms is a great accomplishment. Molecular manufacturing, in contrast, will routinely assemble millions or billions. The basic chemical principles will be the same, but control and reliability will be vastly greater. It is the difference between throwing things together blindly and putting them together with a watchmaker's care.

Technology today doesn't permit thorough control of the structure of matter. Molecular manufacturing will. Today's technologies have given us computers, spacecraft, indoor plumbing, and the other wonders of the modern age. Tomorrow's will do much more, bringing change and choices.

Simple Matter

Today's technology mostly works with matter in a few basic forms: gases, liquids, and solids. Though each form has many varieties, all are comparatively simple.

Gases, as we've seen, consist of molecules ricocheting through space. A volume of gas will push against its walls and, if not walled in, expand without limit. Gases can supply certain raw materials for nanomachines, and nanomachines can be used to remove pollutants from air and turn them into something else. Gases lack structure, so they will remain simple.

Liquids are somewhat like gases, but their molecules cling together to form a coherent blob that won't expand beyond a certain limit. Liquids will be good sources of raw materials for nanomachines because they are denser and can carry a wide range of fuels and raw materials in solution (the pipe in the molecular-processing hall contained liquid). Nanomachines can clean up polluted water as easily as air, removing and transforming noxious molecules. Liquids have more structure than gases, but nanotechnology will have its greatest application to solids.

Solids are diverse. Solid butter consists of molecules stronger than steel, but the molecules cling to one another by the weaker forces of molecular stickiness. A little heat increases thermal vibrations and makes the solid structure disintegrate into a blob of liquid. Butterlike materials would make poor nanomachines. Metals consist of atoms held together by stronger forces, and so they can be structurally stronger and able to withstand higher temperatures.

The forces are not very directional, though, and so planes of metal atoms can slip past one another under pressure; this is why spoons bend, rather than break. This ability to slip makes metals less brittle and easier to shape (with crude technology), but it also weakens them. Only the strongest, hardest, highest-melting-point metals are worth considering as parts of nanomachines.

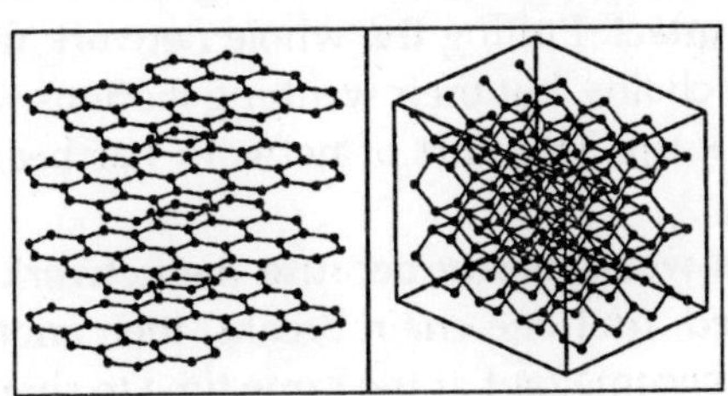

Fig. 1.2 Carbon-Soft and Hard

Diamond consists of carbon atoms held together by strong, directional bonds, like the bonds down the axis of a protein chain. These directional bonds make it hard for planes of atoms to slip past one another, making diamond (and similar materials) very strong indeed–ten to a hundred times stronger than steel. But the planes can't easily slip, so when the material fails, it doesn't bend, it breaks. Tiny cracks can easily grow, making a large object seem weak. Glass is a similar material: glass windows seem weak–and a scratch makes glass far weaker–yet thin, perfect glass fibres are widely used to make composite materials stronger and lighter than steel. Nanotechnology will be able to build with diamond and similar strong materials, making small, flawless fibres and components.

In engineering today, diamond is just beginning to be used. Japan has pioneered a technology for making diamond at low pressure, and a Japanese company sells a speaker with excellent high-frequency response–the speaker cone is reinforced with a light, stiff film of diamond. Diamond is extraordinary stuff, made from cheap materials like natural gas. U.S. companies are scrambling to catch up.

All these materials are simple. More complex structures lead to more complex properties, and begin to give some hint of what molecular manufacturing will mean for materials.

What if you strung carbon atoms in long chains with side-groups, a bit like a protein chain, and linked them into a big three-dimensional mesh? If the chains were kinked so that they couldn't pack tightly, they would coil up and flop around almost like molecules in a liquid, yet the strong bonds would keep the overall mesh intact. Pulling the whole network would tend to straighten the chains, but their writhing motions would tend to coil them back up. This sort of network has been made: it is called rubber.

Rubber is weak mostly because the network is irregular. When stretched, first one chain breaks, then another, because they don't all become taut at the same time to share and divide the load. A more regular mesh would be as soft as rubber at first, but when stretched to the limit would become stronger than steel. Molecular manufacturing could make such stuff.

The natural world contains a host of good materials–cellulose and lignin in wood, stronger-than-steel proteins in spider's silk, hard ceramics in grains of sand, and more. Many products of molecular manufacturing will be designed for great durability, like sand. Others will be designed to fall apart easily for easy recycling, like wood. Some may be designed for uses where they may be thrown away. In this last category, nanotailored biodegradables will shine. With care, almost any sort of product from a shoe to computer-driven nanomachines can be made to last for a good long time, and then unzip fairly rapidly and very thoroughly into molecules and other bits of stuff all of kinds normally found in the soil.

This gives only a hint of what molecular manufacturing will make possible by giving better control of the structure of solid matter. The most impressive applications will not be superstrong structural materials, improved rubber, and simple biodegradable materials: these are uniform, repetitive structures not greatly different from ordinary materials. These materials are "stupid." When pushed, they resist, or they stretch and bounce back. If you shine light on them, they transmit it, reflect it, or absorb it. But molecular manufacturing can do much more. Rather than heaping up simple molecules, it can build materials from trillions of motors, ratchets, light-emitters, and computers.

Muscle is smarter than rubber because it contains molecular machines: it can be told to contract. The products of molecular manufacturing can include materials able to change shape, colour, and other properties on command. When a dust mote can contain a supercomputer, materials can be made smart, medicine can be made sophisticated, and the world will be a different place.

SOURCES OF NANOMATERIALS IN THE ENVIRONMENT

Nanotechnology is a powerful tool for combating cancer and is being put to use in other applications that may reduce pollution, energy consumption, greenhouse gas emissions, and help prevent diseases. NCI's Alliance for Nanotechnology in Cancer is working to ensure that nanotechnologies for cancer applications are developed responsibly.

There is nothing inherently dangerous about being nanosized. Our ability to manipulate objects at the nanoscale has developed relatively recently, but nanoparticles are as old as the earth. Many nanoparticles occur naturally (for example, in volcanic ash and sea spray) and as by-products of human activities since the Stone Age (nanoparticles are in smoke and soot from fire). There are so many ambient incidental nanoparticles, in fact, that one of the challenges of nanoparticle exposure studies is that background incidental nanoparticles are often at order-of-magnitude higher levels than the engineered particles being evaluated.

As with any new technology, the safety of nanotechnology is continuously being tested. The small size, high reactivity, and unique tensile and magnetic properties of nanomaterials—the same properties that drive interest in their biomedical and industrial applications—have raised concerns about implications for the environment, health, and safety (EHS). There has been some as yet unresolved debate recently about the potential toxicity of a specific type of nanomaterial—carbon nanotubes (CNTs)—which has been associated with tissue damage in animal studies. However, the majority of available data indicate that there is nothing uniquely toxic about nanoparticles as a class of materials.

In fact, most engineered nanoparticles are far less toxic than household cleaning products, insecticides used on family pets, and over-the-counter dandruff remedies. Certainly, the nanoparticles used as drug carriers for chemotherapeutics are much less toxic than the drugs they carry and are designed to carry drugs safely to tumors without harming organs and healthy tissue.

To insure that potential risks of nanotechnology are thoroughly evaluated, the NCI Alliance for Nanotechnology in Cancer makes the services of its Nanotechnology Characterization Laboratory (NCL) available to the nanotech and cancer research communities. The NCL, an intramural programme of the Alliance, performs nanomaterial safety and toxicity testing *in vitro* (in the laboratory) and using animal models. The NCL tests are designed to characterize nanomaterials that enter the bloodstream, regardless of route. This testing is just one part of the NCL's cascade of tests to evaluate the physicochemical properties, biocompatibility, and efficacy of nanomaterials intended for cancer therapy and diagnosis. To date, the NCL has evaluated more than 125 different nanoparticles intended for medical applications.

The NCL works closely with the U.S. Food and Drug Administration (FDA) and National Institutes of Standards and Technology (NIST) to devise experiments that are relevant to nanomaterials, to validate these tests on a variety of nanomaterial types, and to disseminate its methods to the nanotech and cancer research communities. The NCL also facilitates the development of voluntary-consensus standards for reliably and pro-actively measuring and monitoring environment, health and safety ramifications of nanotech applications.

Whether actual or perceived, the potential health risks associated with the manufacture and use of nanomaterials must be carefully studied in order to advance our understanding of this field of science and to realise the significant benefits that nanotechnology has to offer society, such as for cancer research, diagnostics, and therapy.

PROPERTIES OF NANOMATERIALS

Over the past decade, nanomaterials have been the subject of enormous interest. These materials, notable for their extremely small feature size, have the potential for wide-ranging industrial, biomedical, and electronic applications. As a result of recent improvement in technologies to see and manipulate these materials, the nanomaterials field has seen a huge increase in funding from private enterprises and government, and academic researchers within the field have formed many partnerships.

Nanomaterials can be metals, ceramics, polymeric materials, or composite materials. Their defining characteristic is a very small feature size in the range of 1-100nanometers (nm). The unit of nanometer derives its prefix nano from a Greek word meaning dwarf or extremely small. One nanometer spans 3-5 atoms lined up in a row. By comparison, the diameter of a human hair is about 5 orders of magnitude larger than a nanoscale particle. Nanomaterials are not simply another step in miniaturization, but a different arena entirely; the nanoworld lies midway between the scale of atomic and quantum phenomena, and the scale of bulk materials. At the nanomaterial level, some material properties are affected by the laws of atomic physics, rather than behaving as traditional bulk materials do.

Although widespread interest in nanomaterials is recent, the concept was raised over 40 years ago. Physicist Richard Feynman delivered a talk in 1959 entitled "There's Plenty of Room at the Bottom", in which he commented that there were no fundamental physical reasons that materials could not be fabricated by maneuvering individual atoms.

Nanomaterials have actually been produced and used by humans for hundreds of years - the beautiful ruby red colour of some glass is due to gold nanoparticles trapped in the glass matrix. The decorative glaze known as luster, found on some medieval pottery, contains metallic spherical nanoparticles dispersed in a complex way in the glaze, which give rise to its special optical properties. The techniques used to produce these materials were considered trade secrets at the time, and are not wholly understood even now.

Development of nanotechnology has been spurred by refinement of tools to see the nanoworld, such as more sophisticated electron microscopy and scanning tunneling microscopy. By 1990, scientists at IBM had managed to position individual xenon atoms on a nickel surface to spell out the company logo, using scanning tunneling microscopy probes, as a demonstration of the extraordinary new technology being developed. In the mid-1980s a new class of material - hollow carbon spheres - was discovered.

These spheres were called buckyballs or fullerenes, in honour of architect and futurist Buckminster Fuller, who designed a geodesic dome with geometry similar to that found on the molecular level in fullerenes. The C_{60} (60 carbon atoms chemically bonded together in a ball-shaped molecule) buckyballs inspired research that led to fabrication of carbon nanofibres, with diameters under 100 nm. In 1991 S. Iijima of

NEC in Japan reported the first observation of carbon nanotubes, which are now produced by a number of companies in commercial quantities. The world market for nanocomposites (one of many types of nanomaterials) grew to millions of pounds by 1999 and is still growing fast.

The variety of nanomaterials is great, and their range of properties and possible applications appear to be enormous, from extraordinarily tiny electronic devices, including miniature batteries, to biomedical uses, and as packaging films, superabsorbants, components of armor, and parts of automobiles. General Motors claims to have the first vehicle to use the materials for exterior automotive applications, in running boards on its mid-size vans. Editors of the journal *Science*profiled work that resulted in molecular-sized electronic circuits as the most important scientific development in 2001. It is clear that researchers are merely on the threshold of understanding and development, and that a great deal of fundamental work remains to be done.

What makes these nanomaterials so different and so intriguing? Their extremely small feature size is of the same scale as the critical size for physical phenomena - for example, the radius of the tip of a crack in a material may be in the range 1-100 nm. The way a crack grows in a larger-scale, bulk material is likely to be different from crack propagation in a nanomaterial where crack and particle size are comparable. Fundamental electronic, magnetic, optical, chemical, and biological processes are also different at this level.

Where proteins are 10-1000 nm in size, and cell walls 1-100 nm thick, their behaviour on encountering a nanomaterial may be quite different from that seen in relation to larger-scale materials. Nanocapsules and nanodevices may present new possibilities for drug delivery, gene therapy, and medical diagnostics.

Surfaces and interfaces are also important in explaining nanomaterial behaviour. In bulk materials, only a relatively small percentage of atoms will be at or near a surface or interface (like a crystal grain boundary). In nanomaterials, the small feature size ensures that many atoms, perhaps half or more in some cases, will be near interfaces. Surface properties such as energy levels, electronic structure, and reactivity can be quite different from interior states, and give rise to quite different material properties.

Let us examine in particular nanocomposites based on polymeric materials, keeping in mind that this is but one small division of nanomaterials. There are several varieties of polymeric nanocomposites, but the most commercially advanced are those that involve dispersion of small amounts of nanoparticles in a polymer matrix.

Those most humble of materials, clays, have been found to impart amazing properties. For example, adding such small amounts as 2% by volume of silicate nanoparticles to a polyimide resin increases the strength by 100%. One should keep in mind, of course, that 2% by volume of very small particles is a great many reinforcing particles. Addition of nanoparticles not only improves the mechanical properties, but also has been shown to improve thermal stability, in some cases allowing use of polymer-matrix nanocomposites an additional 100 degrees Centigrade above the normal service conditions.

Decrease in material flammability has also been studied, an especially important property for transportation applications where choice of material is influenced by safety concerns. Clay/polymer nanocomposites have been considered as matrix materials for fibre-based composites destined for aerospace components. Aircraft and spacecraft components require lightweight materials with high strength and stiffness, among other qualities. Nanocomposites, with their superior thermal resistance, are also attractive for such applications as housings for electronics.

Others have examined the electrical properties of nanocomposites, with an eye to developing new conductive materials. The use of polymer-based nanocomposites has been expanded to anti-corrosion coatings on metals, and thin-film sensors. Theirphotoluminescence and other optical properties are being explored. Polymer-matrix nanocomposites can also be used to package films, an application which exploits their superior barrier properties and low permeability. Although some nanomaterials require rather exotic approaches to synthesis and processing, many polymer-matrix nanocomposites can be prepared quite readily.

Clay/polymer nanocomposites have been made by subjecting a clay such as montmorillonite to ion exchange or other pretreatment, then mixing the particles with polymer melts. There are also a number of other ways to fabricate the materials, including reactive processes involving in situ polymerization.

The low volume fraction of reinforcement particles allows the use of well-established and well-understood processing methods, such as extrusion and injection molding. Ease of processing and forming may be one explanation for the rapidly expanding applications of the materials. Automotive companies, in particular, have quickly adopted nanocomposites in large scale applications, including structural parts of vehicles.

The most energetic research probably concerns carbon nanotubes. Nanoparticles of carbon - rods, fibres, tubes with single walls or double walls, open or closed ends, and straight or spiral forms - have been synthesized in the past 10 years. There is good reason to devote so much effort to them: carbon nanotubes have been shown to have unique properties, stiffness and strength higher than any other material, for example, as well as extraordinary electronic properties.

Carbon nanotubes are reported to be thermally stable in vacuum up to 2800 degrees Centigrade, to have a capacity to carry an electric current a thousand times better than copper wires, and to have twice the thermal conductivity of diamond (which is also a form of carbon). Carbon nanotubes are used as reinforcing particles in nanocomposites, but also have many other potential applications.

They could be the basis for a new era of electronic devices smaller and more powerful than any previously envisioned. Nanocomputers based on carbon nanotubes have already been demonstrated.

NANOMATERIAL STRUCTURE–TOXICITY RELATIONSHIP

As nanotechnology develops into a mature industry the environmental and health effects of its core materials are of increasing importance. A significant challenge for this area of research is that for every class of engineered nanoparticle (nanotubes, metal nanocrystals) there are literally thousands of possible samples with various sizes, surfaces and shapes. This huge parameter space cannot be narrowed by focusing only on commercial materials, as few systems are in commerce at this point. Indeed, most nanotechnology companies are optimizing and evaluating hundreds of material prototypes for possible commercial use. In such a climate, all stakeholders benefit from an understanding of how fundamental nanoparticle characteristics (*e.g.* surface chemistry, size, shape) control their biological effects.

This aim is the overarching objective of this proposal, which stated another way, will provide the first structure-function relationships for nanoparticle toxicology. This information benefits industry in that it will suggest material modifications that may produce systems with minimal environmental and health impact. It also benefits regulators by not only indicating whether information on one nanoparticle type can be used to predict the properties of a related material, but also by setting a framework for evaluating newly developed nanoparticle variants. Finally, a correlation between biological effects and nanoparticle structure will enable the development of chemical methods to alter more toxic nanomaterial species into less toxic materials upon disposal.

To realise these structure-function relationships requires that we develop new analytical tools as well as evaluate material datasets with systematic changes in fundamental properties. Our specific objectives are 1) to expand the characterization of nanoparticle structure in biological media and 2) to characterize the effects of nanoparticles on cell function. This data will be used to test the hypothesis that nanoparticle structure (*e.g.* size and shape) controls directly cytotoxicity. A secondary hypothesis is that of the four major materials parameters in engineered nanoparticles (size, shape, composition and surface) surface will be the most important in governing cellular effects. These hypotheses will be tested in several major classes of nanoparticles.

The rapid proliferation of many different engineered nanomaterials (defined as materials designed and produced to have structural features with at least one dimension of 100 nanometers or less) presents a dilemma to regulators regarding hazard identification. The International Life Sciences Institute Research Foundation/Risk Science Institute convened an expert working group to develop a screening strategy for the hazard identification of engineered nanomaterials. The working group report presents the *elements* of a screening strategy rather than a detailed testing protocol. Based on an evaluation of the limited data currently available, the report presents a broad data gathering strategy applicable to this early stage in the development of a risk assessment process for nanomaterials. Oral, dermal, inhalation, and injection routes of exposure are included recognizing that, depending on use patterns, exposure to nanomaterials may occur by any of these routes. The three key elements of the toxicity screening strategy are: Physicochemical Characteristics, *In Vitro* Assays (cellular and non-cellular), and *In Vivo* Assays.

There is a strong likelihood that biological activity of nanoparticles will depend on physicochemical parameters not routinely considered in toxicity screening studies. Physicochemical properties that may be important in understanding the toxic effects of test materials include particle size and size distribution, agglomeration state, shape, crystal structure, chemical composition, surface area, surface chemistry, surface charge, and porosity.

In vitro techniques allow specific biological and mechanistic pathways to be isolated and tested under controlled conditions, in ways that are not feasible in *in vivo* tests. Tests are suggested for portal-of-entry toxicity for lungs, skin, and the mucosal membranes, and target organ toxicity for endothelium, blood, spleen, liver, nervous system, heart, and kidney. Non-cellular assessment of nanoparticle durability, protein interactions, complement activation, and pro-oxidant activity is also considered.

Tier 1 *in vivo* assays are proposed for pulmonary, oral, skin and injection exposures, and Tier 2 evaluations for pulmonary exposures are also proposed. Tier 1 evaluations include markers of inflammation, oxidant stress, and cell proliferation in portal-of-entry and selected remote organs and tissues. Tier 2 evaluations for pulmonary exposures could include deposition, translocation, and toxicokinetics and biopersistence studies; effects of multiple exposures; potential effects on the reproductive system, placenta, and fetus; alternative animal models; and mechanistic studies.

POTENTIAL FOR HUMAN EXPOSURE

INTRODUCTION

The rapid proliferation of many different engineered nanomaterials presents a dilemma to regulators regarding hazard identification. The screening strategy developed by the International Life Sciences Institute Research Foundation/Risk Science Institute (ILSI RF/RSI) Nanomaterial Toxicity Screening Working Group is an effort to make a significant contribution to the initial hazard identification process for nanomaterial risk assessment.

Engineered nanomaterials are commonly defined as materials designed and produced to have structural features with at least one dimension of 100 nanometers or less. Such materials typically possess nanostructure-dependent properties (*e.g.*, chemical, mechanical, electrical, optical, magnetic, biological), which make them desirable for commercial or medical applications. However, these same properties potentially may lead to nanostructure-dependent biological activity that differs from and is not directly predicted by the bulk properties of the constituent chemicals and compounds. This report outlines the elements of a toxicological screening strategy for nanomaterials as the first step–*i.e.*, hazard identification–in the risk assessment process. Both *in vitro* and *in vivo* methodologies were considered in the development of the screening strategy.

Engineered nanomaterials encompass many forms and are derived from numerous bulk substances. Nanoparticles form a basis for many engineered nanomaterials, and are currently being produced in a wide variety of types for a variety of applications; fullerenes (C_{60} or Bucky Balls), carbon nanotubes (CNT), metal and metal oxide particles, polymer nanoparticles and quantum dots are among the most common.

Engineered nanomaterials are presenting new opportunities to increase the performance of traditional products, and to develop unique new products. "The ability to create unusual nanostructures such as bundles, sheets, and tubes holds promise for new and powerful drug delivery systems, electronic circuits, catalysts, and light-harvesting materials."

Many current efforts are predominantly focused on using relatively simple nanostructured materials such as metal oxide nanoparticles and carbon nanotubes in applications such as high performance materials, energy storage and conversion, self-cleaning surface coatings and stain-resistant textiles. Research into more complex nanomaterials is anticipated to lead to applications such as cellular-level medical diagnostics and treatment and advanced electronics. However, as nanotechnology blurs traditionally rigid boundaries between scientific disciplines, a rapid growth in unanticipated applications is to be expected over the next years and decades.

As new nanotechnology-based materials begin to emerge, it will be essential to have a framework in place within which their potential toxicity can be evaluated, particularly as indicators suggest traditional screening approaches may not be responsive to the nanostructure-related biological activity of these materials.

Several national and international organizations are currently developing standard definitions for comment terms in nanomaterial science including the International Association of Nanotechnology's Nomenclature and Terminology Subcommittee and the American National Standards Institute Nanotechnology Standards Panel (ANSI-NSP). The following key definitions are used throughout this document.

Nanoparticle

A particle with at least one dimension smaller than 100 nm including engineered nanoparticles, ambient ultrafine particles (UFPs) and biological nanoparticles.

Engineered/Manufactured Nanoparticle

A particle engineered or manufactured by humans on the nanoscale with specific physicochemical composition and structure to exploit properties and functions associated with its dimensions. Engineered nanoparticles include particles with a homogeneous composition and structure, compositionally and structurally heterogeneous particles (for instance, particles with core-shell structures) and multi-functional nanoparticles (for instance, 'smart' nanoparticles being developed for medical diagnostics and treatment).

Nanomaterial

A material having a physicochemical structure on a scale greater than typically atomic/ molecular dimensions but less than 100 nm (nanostructure), which exhibits physical, chemical and/or biological characteristics associated with its nanostructure.

Nanostructured Particle

A particle with a physicochemical structure on a scale greater than atomic/molecular dimensions but less than 100 nm, which exhibits physical, chemical and/or biological characteristics associated with its nanostructure. A nanostructured particle may be much larger than 100 nm. For example, agglomerates of TiO_2 nanoparticles that are significantly larger than 100 nm in diameter may have a biological activity determined by their nanoscale sub-structure.

Other examples include zeolites, meso-porous materials and multifunctional particulate probes.

Agglomerate/Aggregate

The terms "agglomerate" and "aggregate" are used differently and even interchangeably in different fields. In the context of this report, the term "agglomerate" is used exclusively to describe a collection of particles that are held together by both weak and strong forces, including van der Waals and electrostatic forces, and sintered bonds. In this document, the term is used interchangeably with 'aggregate'. However, the importance of understanding how the binding forces of an agglomerate affect the dispersibility of the component particles under different conditions–in essence how easily the agglomerate de-agglomerates–is noted.

Nanoporous Material

A material with particles that are larger than 100 nm may have significant structuring on the nanometer size scale, thereby providing properties based upon this smaller structuring that may be toxicologically relevant (*e.g.*, dramatically increased surface area as compared to the bulk). Nanoporous materials, such as zeolites, are a significant class of materials which have porosity in the sub-100 nm size range but whose primary particles may be large.

OBJECTIVES

The objective of the ILSI RSI Nanomaterial Toxicity Screening Working Group, which was convened in February 2005, was to identify the key elements of a toxicity screening strategy for engineered nanomaterials. The group considered potential effects of exposure to nanomaterials by inhalation, dermal, oral, and injection routes; discussed how mechanisms of nanoparticle toxicity may differ from those exhibited by larger particles of the same chemical; and identified significant data needs for designing a robust screening strategy. The elements of a screening strategy for nanomaterials presented by the

Nanomaterial Toxicity Screening Working Group include an evaluation of the physicochemical characteristics and dose metrics; acellular assays; *in vitro* assays for lung, skin, and mucosal membranes; and *in vivo*assays for lung, skin, oral, and injection exposures.

This project was funded by the U.S. Environmental Protection Agency Office of Pollution Prevention and Toxics through a cooperative agreement with the ILSI Research Foundation/Risk Science Institute. It was an outgrowth of another project under the same cooperative agreement that proposed strategies for short-term toxicity testing of fibres. Among the principal conclusions of the latter project were the importance of the physicochemical characterization of the fibres, the value of subchronic (1–3 month) rat inhalation exposure studies, and the typically key role in fibre toxicity of biopersistence of inhaled fibres in the lung and of chronic inflammation leading to cell proliferation and interstitial fibrosis.

LITERATURE SURVEY

The potential for human and ecological toxicity associated with nanomaterials and ultrafine particles is a growing area of investigation as more nanomaterials and products are developed and brought into commercial use. To date, few nanotoxicology studies have addressed the effects of nanomaterials in a variety of organisms and environments. However, the existing research raises some concerns about the safety of nanomaterials and has led to increased interest in studying the toxicity of nanomaterials for use in risk assessment and protection of human health and the environment.

A new field of nanotoxicology has been developed to investigate the possibility of harmful effects due to exposure to nanomaterials. Nanotoxicology also encompasses the proper characterization of nanomaterials used in toxicity studies. Characterization has been important in differentiating between naturally occurring forms of nanomaterials, nano-scale byproducts of natural or chemical processes, and manufactured (engineered) nanomaterials. Because of the wide differences in properties among nanomaterials, each of these types of nanoparticles can elicit its own unique biological or ecological responses. As a result, different types of nanomaterials must be categorized, characterized, and studied separately, although certain concepts of nanotoxicology based on the small size, likely apply to all nanomaterials.

As materials reach the nanoscale, they often no longer display the same reactivity as the bulk compound. For example, even a traditionally inert bulk compound, such as gold, may elicit a biological response when it is introduced as a nanomaterial. New approaches for testing and new ways of thinking about current materials are necessary to provide safe workplaces, products, and environments as the manufacturing of nanomaterials and products increases and, as a result, exposure to nanomaterials increases. The diverse routes of exposure, including inhalation, dermal uptake, ingestion, and injection, can present unique toxicological outcomes that vary with the physicochemical properties of the nanoparticles in question.

The earliest studies investigating the toxicity of nanoparticles focused on atmospheric exposure of humans and environmentally relevant species to heterogeneous mixtures of environmentally produced ultrafine particulate matter (having a diameter <100 nm). These studies examined pulmonary toxicity associated with particulate matter deposition in the respiratory tract of target organisms. Epidemiological assessments of the effects of urban air pollution exposure focusing on particulate matter produced as a byproduct of combustion events, such as automobile exhaust and other sources of urban air pollution, showed a link in test populations between morbidity and mortality and the amount of particulate matter. Some researchers have found an increased risk of childhood and adult asthma correlated to environmental exposure to ultrafine particulate matter in urban air. However, other research does not indicate the same correlation.

Laboratory-based studies have investigated the effects of a large range of ultrafine materials through *in vivo* exposures using various animal models as well as cell-culture-based *in vitro*experiments. To date, animal studies routinely show an increase in pulmonary inflammation, oxidative stress, and distal organ involvement upon respiratory exposure to inhaled or implanted ultrafine particulate matter.

Tissue and cell culture analysis have also supported the physiological response seen in whole animal models and yielded data pointing to an increased incidence of oxidative stress, inflammatory cytokine production, and apoptosis in response to exposure to ultrafine particles. These studies have also yielded information on gene expression and cell signaling pathways that are activated in response to exposure to a variety of ultrafine particle species ranging from carbon-based combustion products to transition metals. Polytetrafluoroethylene fumes in indoor air pollution are nano-sized particles, highly toxic to rats. They elicit a severe inflammatory response at low inhaled particle mass concentrations, suggestive of an oxidative injury.

In contrast to the heterogeneous ultrafine materials produced incidentally by combustion or friction, manufactured nanomaterials can be synthesized in highly homogenous forms of desired sizes and shapes (*e.g.*, spheres, fibres, tubes, rings, planes). Limited research on manufactured nanomaterials has investigated the interrelationship between the size, shape, and dose of a material and its biological effects, and whether a unique toxicological profile may be observed for these different properties within biological models.

Typically, the biological activity of particles increases as the particle size decreases. Smaller particles occupy less volume, resulting in a larger number of particles with a greater surface area per unit mass and increased potential for biological interaction. Recent studies have begun to categorize the biological response elicited by various nanomaterials both in the ecosystem and in mammalian systems. Although most current research has focused on the effect of nanomaterials in mammalian systems, some recent studies have shown the potential of nanomaterials to elicit a phytotoxic response in the ecosystem. In the case of alumina nanoparticles, one of the US market leaders for nano-sized materials, 99.6% pure nanoparticles with an average particle size of 13 nm were shown to cause root growth inhibition in five plant species.

Toxicological studies of fibrous and tubular nanostructures have shown that at extremely high doses these materials are associated with fibrotic lung responses and result in inflammation and an increased risk of carcinogenesis. Single-walled carbon nanotubes (SWCNT) have been shown to inhibit the proliferation of kidney cells in cell culture by inducing cell apoptosis and decreasing cellular adhesive ability. In addition, they cause inflammation in the lung upon instillation. Multi-walled carbon nanotubes (MWCNT) are persistent in the deep lung after inhalation and, once there, are able to induce both inflammatory and fibrotic reactions.

Dermal exposure to MWCNT has been modeled through cell culture and points to the nanoparticles' ability to localize within and initiate an irritation response in target epithelial cells. Proteomic analysis conducted in human epidermal keratinocytes exposed to MWCNT showed both increased and decreased expression of many proteins relative to controls. These protein alterations suggested dysregulation of intermediate filament expression, cell cycle inhibition, altered vesicular trafficking/exocytosis and membrane scaffold protein down-regulation. In addition, gene expression profiling was conducted on human epidermal keratinocytes exposed to SWCNT that showed a similar profile to alpha-quartz or silica. Also, genes not previously associated with these particulates before from structural protein and cytokine families were significantly expressed. Dosing keratinocytes and bronchial epithelial cells *in vitro* with SWCNT has been shown to result in increases in markers of oxidative stress.

Charge properties and the ability of carbon nanoparticles to affect the integrity of the blood-brain barrier as well as exhibit chemical effects within the brain have also been studied. Nanoparticles can overcome this physical and electrostatic barrier to the brain. In addition, high concentrations of anionic nanoparticles and cationic nanoparticles are capable of disrupting the integrity of the blood-brain barrier. The brain uptake rates of anionic nanoparticles at lower concentrations were greater than those of neutral or cationic formulations at the same concentrations. This work suggests that neutral nanoparticles and low concentration anionic nanoparticles can serve as carrier molecules providing chemicals direct access to the brain and that cationic nanoparticles have an immediate toxic effect at the blood-brain barrier.

Tests with uncoated, water soluble, colloidal C_{60} fullerenes have shown that redox-active, lipophilic carbon nanoparticles are capable of producing oxidative damage in the brains of aquatic species. The bactericidal potential of C_{60} fullerenes was also observed in these experiments. This property of fullerenes has possible ecological ramifications and is being explored as a potential source of new antimicrobial agents.

Oxidative stress as a common mechanism for cell damage induced by nano- and ultrafine particles is well documented; fullerenes are model compounds for producing superoxide. A wide range of nanomaterial species have been shown to create reactive oxygen species both *in vivo*and *in vitro*. Species which have been shown to induce free radical damage include the C_{60}fullerenes, quantum dots, and carbon nanotubes. Nanoparticles of various sizes and chemical compositions are able to preferentially localize in mitochondria where they induce major structural damage and can contribute to oxidative stress.

Quantum dots (QDs) such as CdSe QDs have been introduced as new fluorophores for use in bioimaging. When conjugated with antibodies, they are used for immunostaining due to their bright, photostable fluorescence.

To date, there is not sufficient analysis of the toxicity of quantum dots in the literature, but some current studies point to issues of concern when these nanomaterials are introduced into biological systems. Recently published research indicates that there is a range of concentrations where quantum dots used in bioimaging have the potential to decrease cell viability, or even cause cell death, thus suggesting that further toxicological evaluation is urgently needed. While it is well known that bulk cadmium selenide (CdSe) is cytotoxic, it has been suggested that CdSe quantum dots are cytocompatible, and safe for use in whole animal studies. This postulate is based in part on the use of protecting groups around the CdSe core of the quantum dot. These coatings have been shown to be protective, but their long-term stability has not been evaluated thoroughly. Recent studies exploring the cytotoxicity of CdSe-core quantum dots in primary hepatocytes as a liver model found that these quantum dots were acutely toxic under certain conditions. The cytotoxicity correlates with the liberation of free Cd^{2+} ions due to deterioration of the CdSe lattice. These data suggest that quantum dots can be rendered nontoxic initially for *in vivo* use when appropriately coated. However, the research also highlights the need to further explore the long-term stability of the coatings used, both *in vivo* and exposed to environmental conditions.

The range of approaches and methods used to reach conclusions regarding the effects of manufactured nanomaterials and ultrafine particles has led to different results. This inconsistency indicates a need for standardized tests in order to get comparable results in screening nanomaterials for potential adverse effects. As the field of nanotoxicology continues to grow, standard toxicology tests will aid those entering the field and allow for better comparisons and conclusions in determining the toxic effects of nanomaterials.

2

Characterization of Nanomaterials

PHYSICOCHEMICAL CHARACTERIZATION

Unlike gases, liquids and many solid materials, the desirable properties of engineered nanomaterials closely depend on size, shape and structure (both physically and chemically) at the nanoscale. Similarly, there is a strong likelihood that biological activity will depend on physicochemical parameters not usually considered in toxicity screening studies.

Although quantitative toxicity studies on engineered nanomaterials are still relatively sparse, published data on fullerenes, single walled carbon nanotubes, nanoscale metal oxides such as TiO_2 and nanometer-diameter low solubility particles, support the need to carefully consider how nanomaterials are characterized when evaluating potential biological activity. Respirable fibres present perhaps the closest analogy to a material that is not fully characterized by mass and chemical composition alone. However, the diversity and complexity of nanomaterials suggests that the level of characterization appropriate to toxicity screening tests will be commensurately more sophisticated.

Until the mechanistic associations between nanomaterial characteristics and toxicity are more fully understood, it will be necessary to ensure that all nanomaterial characteristics that are potentially significant are measured or can be derived in toxicity screening tests. In particular, in as far as it is possible; it is desirable to collect sufficient information to allow retrospective interpretation of toxicity data in the light of new findings.

In this context, identifying a set of characterization criteria for nanomaterial toxicity screening presents a significant challenge. Clearly, the ideal of characterizing every possible aspect of a test material, while laudable, is impractical. In this document, we have therefore focused on the context under which characterization takes place and the minimum set of characterization parameters we consider essential within that context. Essential parameters have been supplemented with those considered desirable and those considered of interest but optional within a screening study. The two overarching characterization contexts discussed are human exposure studies and *in vitro/in vivo* studies. In the case of the latter, we consider material characterization after administration, characterization at the point of administration and characterization of the bulk material as produced or supplied. The relative importance of characterizing dose against different physical metrics during inhalation exposures is also discussed. Recommendations are subsequently made on physicochemical characterizations for nanomaterial toxicity screening tests and characterization methods capable of providing the recommended information.

Framework for Material Characterization

Material characterization for toxicity screening studies is most appropriately considered in the context of the studies being undertaken. Requirements for *in vitro* and *in vivo* screening studies will differ according to the material delivery route or method. Additionally, understanding human exposures in the context of developing appropriate screening studies will present a further set of characterization requirements.

Four screening study contexts are proposed, and characterization recommendations are developed within these contexts:

- Human exposure characterization

- Characterization of material following administration
- Characterization of administered material
- Characterization of as-produced or supplied material

Human Exposure Characterization

Where exposure to a specific material is known to occur or is anticipated, exposure studies are desirable in developing and selecting appropriate toxicity screening tests. At present engineered nanomaterials are predominantly at the research or pre-production stage, and there are relatively few environments where exposures are known to occur.

However, if commercialization of products using nanomaterials develops as anticipated, the potential for exposure is likely to increase dramatically over the coming decade. Therefore, estimates of future use and potential human exposures should be considered in the development of toxicity screening.

Nanomaterial Characterization after Administration

Characterizing delivered nanomaterial after administration in a test system or model provides the highest quality of data on dose and material properties that are related to observed responses, but this is limited by current methodological capabilities. Characterization after administration is particularly advantageous where the possibility of physicochemical changes in the material before and after administration exists.

Examples of potential changes include aggregation state, physisorption or chemisorption of biomolecules and biochemically-induced changes in surface chemistry. In addition, possible physicochemical changes as a result of nanomaterial interactions with the surrounding biological systems such as rapid dissolution of water- or lipid soluble fractions of the nanomaterial need to be carefully considered.

While characterization after administration is considered an ideal to work towards, it is recognized that in many cases, characterization at the point of administration will be a more realistic and feasible option. It is also recognized that in many cases, characterization at the point of administration will be essential for the intercomparison of studies, irrespective of whether characterization after administration is carried out.

Characterization of Administered Material

Characterization of administered material in toxicity screening studies is fundamental. This approach addresses potential physicochemical changes between the bulk material and the administered material (such as agglomeration state) and allows more robust causal associations between the material and observed responses to be developed.

However, given the strong sensitivity of many nanomaterial properties to their local environment, it should be noted that biologically relevant changes in the physicochemical nature of a nanomaterial between administration and deposition may have a significant impact on observed responses in some instances.

Characterization of as-Produced or Supplied Material

Characterization of nanomaterials as-produced or as-supplied represents the most direct approach to obtaining physicochemical information and may provide useful baseline data on the material under test. Most engineered nanomaterials have a functionality based on their physicochemistry. It is therefore likely that information of relevance to toxicity screening studies will be available from suppliers or producers in many cases.

However, due to the current lack of accepted nanomaterial characterization standards, it is strongly recommended that wherever possible, independent characterization of test nanomaterials be conducted.

Characterization of supplied nanomaterial may not appropriately represent physicochemical properties of the material during or following administration. For this reason exclusive reliance on this approach is discouraged, and is only recommended where characterization of material during or after administration is clearly not feasible.

KEY CHARACTERISTICS

Previous studies of asbestos and other fibres have shown that the dimension, durability and dose (the three D's) of fibrous particles are key parameters with respect to their pathogenicity. In general, fibres with a smaller diameter will penetrate deeper in the lungs. Long fibres (longer than the diameter of alveolar macrophages) stimulate macrophages to release inflammatory mediators and will only be cleared slowly.

In addition to fibre length, chemical factors play an important role in fibre durability and biopersistence; fibres with high alkali or alkali earth oxide contents and low contents of Al_2O_3, Fe_2O_3, TiO_2 tend to have low durability and hence low biopersistence. On the other hand, studies of mineral particles have demonstrated that the toxic and carcinogenic effects are, in some cases, related to the surface area of inhaled particles and their surface activity. Particle surface characteristics are considered to be key factors in the generation of free radicals and reactive oxygen species formation and in the development of fibrosis and cancer by quartz (crystallized silica).

The unusual properties of nanomaterials are predominantly associated with their nanometer-scale structure, size and structure-dependent electronic configurations and an extremely large surface-to-volume ratio relative to bulk materials. Particles in the nanosize range can deposit in all regions of the respiratory tract including the distal lungs. Due to their small size, nanoparticles may pass into cells directly through the cell membrane or penetrate between or through cells and translocate to other parts of the body. Limited data have suggested possible translocation of inhaled nanoparticles to the nervous system and other organs/tissues.

The size of nanoparticles alone may not be the critical factor determining their toxicity; the overall number and thus the total surface area may also be important. As a particle decreases in size, the surface area increases (per unit mass only; if you normalize to number of particles, the surface area decreases) and a greater proportion of atoms/molecules are found at the surface compared to those inside. Thus, nanoparticles have a much larger surface area per unit mass compared with larger particles. The increase in the surface-to-volume ratio results in the increase of the particle surface energy which may render them more biologically reactive.

Nano-scale materials are known to have various shapes and structures such as spheres, needles, tubes, plates, etc. Nanoporous materials are materials with defined pore-sizes in the nanometer range. The effects of the shape on the toxicity of nanomaterials are unknown. The shape of nanomaterials may have effects on the kinetics of deposition and absorption in the body. The results of a recent *in vitro* cytotoxicity study appear to suggest that single-wall nanotubes are more toxic than multi-wall nanotubes.

Chemical composition is another important parameter for the characterization of nanomaterials, which comprise nearly all substance classes, *e.g.*, metal/metal oxides, compounds, polymers as well as biomolecules. Some nanomaterials can also be a combination of the above components in core-shell or other complex structures. Dependent on the particle surface chemistry, reactive groups on a particle surface will certainly modify the biological effects.

Under ambient conditions, some nanoparticles can form aggregates or agglomerates. These agglomerates have various forms, from dendritic structure to chain or spherical structures. To maintain the characteristics of nanoparticles, they are often stabilized with coatings or derivative surface to prevent agglomeration. The properties of nanoparticles can be significantly altered by surface modification and the distribution of nanoparticles in the body strongly depends upon the surface characteristics. Changes of surface properties by coating of nanoparticles to prevent aggregation or agglomeration with different types and concentrations of surfactants have been shown to change their body distribution and the effects on the biological systems significantly.

Therefore, it is recommended that the following physicochemical properties of the test materials should be characterized:

- Size distribution
- Agglomeration state
- Shape
- Crystal structure
- Chemical composition–including spatially averaged (bulk) and spatially resolved heterogeneous composition
- Surface area
- Surface chemistry
- Surface charge
- Porosity

DOSE METRICS

In any toxicity screening study, careful consideration should be given to the metric used to quantify dose. Although response may be associated with a wide range of physicochemical characteristics, measuring dose against a physical metric of mass, surface area or particle number for a well-characterized material will enable quantitative interpretation of data. Appropriate selection of the dose metric will depend on the hypothesized parameter most closely associated with anticipated response or the metric which may be most accurately measured.

It is strongly recommended that in all cases, sufficient information is collected to enable dose against *all three primary physical metrics* to be derived. This may be achieved where the relationships between nanomaterial mass, surface area and particle number concentration are known, or where measurements of particle size distribution are made that enable derivation of all three dose metrics.

Where nanomaterials are administered in a liquid medium, such as in the technique of intratracheal instillation or pharyngeal aspiration, the nature and amount of material within the suspension should be fully characterized before delivery in terms of number, surface area and mass concentration.

Inhalation studies present additional challenges of measuring dose over time, and require both on-line (time resolved) and off-line analysis.

Off-line mass concentration measurements using filter-based methods offer continuity with standard inhalation studies and are recommended as an essential component of inhalation nanomaterial screening tests. Likewise, on-line mass concentration measurements are recommended as an essential component of inhalation studies.

Gravimetric and/or chemical analysis of filter samples will provide the most accurate characterization of exposure in many cases when compared to off-line surface area and number concentration analyses. With appropriate additional information, such measurements may be used to calculate aerosol surface area or number concentration. However, the diameter cubed relationship between particle size and mass can lead to large errors when transforming from mass to number concentration if the size distribution is broad or there are small numbers of excessively large particles present.

On-line mass-concentration measurements using instruments such as the Tapered Element Oscillating Microbalance (TEOM) potentially offer high precision and good accuracy, although they are susceptible to errors where the sampled aerosol contains volatile components. On-line photometric mass concentration methods are generally good for monitoring the temporal stability of aerosol and providing a real-time indication of mass exposure, although they are relatively insensitive to particles smaller than approximately 0.5 μm in diameter. However, in general more appropriate methods should be used for providing real-time measurements of number and surface-area exposure.

Aerosol size distribution measurements enable reasonably good calculation of exposure against all three physical metrics, if parameters such as particle shape and density are known. Off-line size distribution measurement methods such as Transmission Electron Microscopy (TEM) analysis offer detailed information on this distribution but are extremely time consuming, and frequently limited by the collection techniques and, in the case of TEM analysis, inference of 3-dimensional structure from 2-dimensional images. On-line size measurement techniques such as Differential Mobility Analysis are capable of measuring aerosol size distribution with a time resolution of tens of seconds. Aerosol number concentration between given particle diameters is easily derived from aerosol size distribution measurements, although interpretation of such data in terms of mass or surface-area dose requires additional information on particle characteristics such as shape and density.

It is recommended that for each nanoparticle type, size distribution measurement techniques be validated against TEM analysis. Off-line surface area characterization is possible using isothermal gas-adsorption, although techniques suited to filter samples need to be employed. There is also some possibility that the surface area of the collected material will differ from that of the airborne material due to compaction and surface occlusion. However, the extent to which this may occur is not well understood.

Published studies have shown a good correlation between off-line surface area measurement and biological response, suggesting that errors associated with collection and subsequent analysis can be small. This holds particularly for insoluble particles; ideally surface area measurements would be required on the insoluble core of a nanomaterial after its water-and/or lipid soluble compounds have been dissolved from the particle surface. Aerosol diffusion charging has been shown to provide a measure of surface area on-line where the charging rate is low, and a small number of aerosol diffusion chargers are commercially available.

These devices have been shown to measure aerosol surface area well for particles smaller than 100 nm in diameter. At larger diameters, measured surface area progressively underestimates aerosol surface area. In particular, the surface area of porous particle structures as well as that of highly aggregated particles will generally not be determined. Data have been published on a particular aerosol diffusion charger indicating that it provides a measure of aerosol surface area dose in the lungs, as opposed to aerosol surface area exposure. While on-line aerosol surface area measurements are desirable during inhalation exposure studies, uncertainties associated with current techniques suggest caution when interpreting such measurements.

Number concentration may be measured on-line with relative ease using instruments such as Condensation Particle Counters. Although it is not clear how biologically relevant number concentration is as a dose metric, the ease with which such measurements are made and their value in tracking temporal changes in exposure lead to their being recommended as essential in inhalation studies.

CHARACTERIZATION PRIORITIZATION

In developing recommendations on material characterizations for nanomaterial toxicity screening studies, three specific factors have been taken into consideration: the context within which a material is being evaluated, the importance of measuring a specific parameter within that context, and the feasibility of measuring the parameter within a specific context.

In addition, recommendations have been made on recording information on nanomaterial production, preparation, storage, heterogeneity, and agglomeration state. To enable retrospective interpretation of toxicity data and replication of tests, it is strongly recommended that all information on the production and processing of nanomaterials be recorded. Fully documenting storage time and conditions (including temperature, humidity, exposure to light and atmosphere composition) is essential, as physicochemical changes may take place over time.

If possible, the physicochemical stability of samples over time should be demonstrated. Where a test material is a heterogeneous mixture of different components, information is required on the relative abundance of the different components, and whether associations in the bulk material are maintained in the administered material, or whether different components are preferentially administered with specific delivery mechanisms. The agglomeration state of a nanomaterial during and following administration may have a significant impact on its biological activity.

Agglomeration state at different structure scales should be characterized, including primary (primary particles), secondary (primary particle agglomerates and self-assembled structures) and tertiary (assemblies of secondary structures) scales. Ideally, agglomeration state in the biological environment following administration should be evaluated. If possible, some insight into the binding forces within agglomerates (*e.g.* relatively weak van der Waals forces or relatively strong sintered bonds) should be obtained. Material agglomeration or de-agglomeration in different liquid media should also be investigated where possible.

Characterization of material as administered is recommended as the highest priority, supplemented by characterization after *in vitro* or *in vivo* administration where possible, and followed in order of preference by characterization of the material as produced or supplied.

ANALYSIS METHODS

Techniques have been categorized with respect to their applicability to specific material characteristics. In general, the table is self-explanatory, and further information on each technique can be obtained from a wide range of sources. A number of techniques are only suitable for materials in certain forms, or specific classes of materials.

For instance, while Transmission Electron Microscopy is capable of providing a wealth of information on nanoparticles and is considered a gold standard for evaluating particle size distribution and shape, dry (or in the case of cryo-TEM, frozen liquid-encapsulated) well-dispersed samples that are sufficiently robust to withstand high vacuums are required. Similarly, techniques such as Infrared (IR) spectroscopy are particularly sensitive to surface organic compounds, but are less useful for quantifying inorganic surface chemistry. Applicability of a range of analytical techniques to providing specific physicochemical information on engineered nanomaterials, in the context of toxicity screening studies

Given the wide range of analytical techniques available in many disciplines associated with nanotechnology, multidisciplinary collaborations with research and analysis groups offering state of the art nanomaterial characterization capabilities are strongly recommended when carrying out nanomaterial toxicity screening studies.

RESEARCH GAPS

- The development of viable *in vivo* nanomaterial (including nanoparticles) detection techniques.
- The development and production of inexpensive real-time monitoring instruments and methods for aerosol mass concentration (low concentrations, nanoscale particles), surface area concentration and size distribution.
- The development of standardized, well characterized nanomaterial samples.
- The development of radio-labeled nanomaterial samples, and samples that can be tracked and detected through neutron-activation.
- The development of more advanced surface chemistry characterization techniques, in particular techniques capable of detecting and speciating biological molecules on the surface of nanoparticles and nanomaterials.
- The development of electron microscopy techniques for biologically-relevant nanoscale analysis.

RECOMMENDATIONS

- All nanomaterial physicochemical characteristics that are potentially significant should be measured or be derivable in toxicity screening tests.
- Characterization of nanomaterial as administered is strongly recommended, supplemented by characte-

rization following administration where it is technically feasible and practicable. Characterization of the bulk material as-produced or supplied to the exclusion of the above is not recommended, except where more appropriate measurements are not feasible.

- It is recommended that independent characterizations of nanomaterials (beyond information provided by producers and suppliers) are carried out where possible.
- It is recommended that the following physicochemical properties of nanomaterials should be characterized in the context of toxicity screening tests: Particle size distribution, agglomeration state, particle shape, crystal structure, chemical composition (bulk and spatial), surface area, surface chemistry, surface charge, and porosity.
- It is recommended that in all cases, sufficient information be collected to enable derivation of the delivered dose against all three primary physical metrics (number, surface area and mass concentration).
- Off-line mass concentration measurements using filter-based methods are recommended as an essential component of inhalation nanomaterial screening tests. In addition, off-line measurement of aerosol size distribution is recommended.
- On-line mass concentration and number measurements are recommended as an essential component of inhalation studies.
- Multidisciplinary collaborations between research and analysis groups offering state of the art nanomaterial characterization capabilities are strongly recommended.
- It is recommended that information on nanomaterial production, preparation, storage, heterogeneity and agglomeration state be recorded for all nanomaterial toxicity screening studies.

- It is recommended that nanomaterial preparation methods are fully documented, including the selection of appropriate dispersion media, methods of dispersion in the medium and agglomeration state within the medium. Specific preparation techniques are not recommended, as these will depend on the material and test protocols being used. However, caution is advised when using ultrasonic agitation to disperse materials, as at high energies the method may be sufficiently aggressive to alter the material characteristics.

TESTING METHODS

INTRODUCTION

Before considering the application of specific *in vitro* testing methods to the assessment of the toxicity of nanomaterials, there are several generic issues that should be noted.

Advantages and Disadvantages

In general *in vitro* techniques are seen as an important adjunct to *in vivo* studies. These studies allow specific biological pathways to be tested under controlled conditions, as well as isolation of pathways that is not feasible *in vivo*; *e.g.*, it is difficult to discriminate *in vivo* whether complement activation has a role in any pro-inflammatory effects of particles. The complement system can be isolated *in vitro*, and its potential role investigated. There are, of course, well-documented problems with *in vitro* approaches, including lack of validation against *in vivo* adverse effects, dosimetry mismatch, over-simplicity, non-involvement of the complete inflammatory response, etc.

Control Particles

It is important, in view of the above, that adequate positive and negative control particles are included in all experiments. This at least allows the test particle to be bench-marked against particles of known toxicity. These can include standard crystalline silica (quartz; *e.g*, Min-U-Sil or DQ12) as a known cytotoxic particle and fine TiO2 as an inert particle.

Expression of Dose

Toxicity and other responses should be expressed in relation to a range of dose metrics depending on the material and the dose metric data that are available.

Adsorption of Proteins by Nanoparticles

The large surface area of nanoparticles means that they are capable of adsorbing proteins. Nanoparticles of various types have been reported to adsorb key proteins such as albumin, fibronectin and TGF-β. This may confound endpoints that rely on the measurement of a protein as the protein may be produced but may also remove from the supernatant onto the nanoparticle surface by adsorption, providing a false-negative.

The *in vitro* tests that are presented will be divided into portal of entry toxicity and target organ toxicity. The potential target cells and associated appropriate endpoints will be described. Finally, research gaps and recommendations will be identified.

PORTALS OF ENTRY

Lungs

The lungs represent a potential target for any airborne particles, and many *in vitro* models for the lung exist. Particles deposit on the airway or alveolar epithelium and encounter mucus or epithelial lining fluid. They may then interact with macrophages, which may result in their clearance, or they may enter the interstitium where they may make contact with fibroblasts and endothelial cells or cells of the immune system.

The Epithelium

The epithelium is the first barrier that confronts particles that deposit in either the conducting airways or the alveolar region. Therefore, both bronchial and alveolar epithelial cells should be considered as target cells for *in vitro* studies. Endpoints for detecting nanoparticle effects could include toxicity measurements, such as LDH release, for necrosis or various cytokine expression (IL-8, MCP-1 etc), and activation of inflammation-related transcription factors such as NF-κB and AP-1.

Oxidative and nitrosative stress are dominant mechanistic hypotheses for cell damage and activation caused by pathogenic particles. These can be monitored by measuring oxidative stress using dichlorofluorescein or oxidized glutathione as endpoints and nitrosated proteins as a measure of active nitrogen species. Responses to particle-induced oxidative/nitrosative stress can include up-regulation of anti-oxidant genes such as superoxide dismutase and glutathione peroxidase, and so these can also be measured. Proliferative effects of nanoparticles can be assessed using a variety of assays including bromo-deoxyuridine incorporation.

If cancer is an endpoint that is under consideration, then direct measures of genotoxicity can be quantified by methods that include COMET assay and 8-hydroxy-deoxyguanosine measurement. The translocation of nanoparticles across the epithelium could be an important discriminator of harmfulness and, although there are few publications specifically addressing transfer of particles across the epithelium *in vitro*, these should be developed and could contribute to understanding the factors that regulate translocation.

Macrophages

Macrophages play a key role in the cellular response to particles that deposit in the lungs. Macrophages could be affected by nanoparticles in various ways that can be studied *in vitro*through a variety of assays. Cellular cytotoxicity could be measured using conventional methods, such as lactate dehydrogenase release. Macrophage activation occurs following phagocytosis of a number of pathogenic particles leading to release of cytokines (tumour necrosis factor alpha (TNFα), interleukin-6 (IL-6) etc) and nuclear transfer of inflammation-related transcription factors nuclear factor kappa B (NF-κB) and activator protein 1(AP-1).

Macrophages undergo an oxidative burst (OB) on phagocytosis of particles and the extent of this in response to nanoparticles could be investigated. Nitric oxide (NO) may also be produced, in response to particles and in the presence of superoxide radical peroxynitrite, a highly toxic species, can be produced.

If the OB or NO production is exaggerated, there could be 'bystander' injury to epithelial cells whilst diminished OB/NO production could mean impaired microbicidal activity that allows infection.

Another key macrophage function reported to be impaired by nanoparticles is phagocytosis, and so the effect of test nanoparticles on this function could be considered. The cytoskeleton is key to normal cell functioning and could be targeted by nanoparticles and so could be investigated.

Endothelial Cells

Although these are found in the lungs, they are considered a part of the cardiovascular system and are dealt with below.

Fibroblasts

Fibroblasts are found in the interstitium and are liable to be affected by any particle that gains access to this site. At least two important modes of response could be activated by nanoparticle/fibroblast interactions and both modes constitute relevant endpoints for *in vitro*testing:

- Pro-inflammatory effects, measured by cytokine/ chemokine gene expression (TNFα; etc); or
- Fibrogenic responses activated either by direct stimulation of fibroblast growth or extra-cellular matrix secretion by the nanoparticle, or by autocrine stimulation following nanoparticle-stimulated release from the fibroblasts of growth factors such as transforming growth factor beta and platelet-derived growth factor.

The Immune System

Immunopathological effects could be envisaged if particles interact with lymphocytes, or as a consequence of their predilection for entering the interstitium, they modulate dendritic cell function.

The effects of nanoparticles on immunological functions including antigen presentation by macrophages and dendritic cells and the subsequent effects on immune responses *in vitro* are relevant endpoints and appropriate tests should be designed.

Co-Cultures

In addition to monocultures of lung cells, co-cultures such as epithelial cells/macrophages or epithelial cells/endothelial cells may more closely represent the *in vivo* situation, and so such studies are encouraged.

Lung Slices

Methodology to culture whole lung tissue slices is available, such that multiple pulmonary cell types can be exposed *in vitro* in the same configuration as they occur *in vivo*.

Cell Lines vs. Freshly-Derived Cells

If possible, freshly-derived primary cells should be used. Where cell lines are used, these should preferably not be cancer cells. Where cancer cells are used, the endpoint response under study should be carefully compared to non-cancer cells to ensure that, for that endpoint, the fact that the cell is a cancer cell does not greatly modify the response compared to a non-cancer cell.

Whole Heart-Lung Preparation

The Langendorff heart-lung preparation may provide the opportunity to study the behaviour of nanoparticles under highly controlled conditions. In this model the exsanguinated heart and lungs are maintained by perfusion and so transport between the lungs and the vascular space can be studied in the absence of blood.

Skin

Skin or the integument is the largest organ of the body and is unique because it is a potential route for exposure to nanoparticles during their manufacture and also provides an environment within the avascular epidermis where particles could potentially lodge and not be susceptible to removal by phagocytosis.

What are the toxicological consequences of "dirty" nanoparticles (catalyst residue) becoming lodged in the epidermis? In fact, it is this relative biological isolation in the lipid domains of the epidermis that has allowed for the delivery of drugs to the skin using lipid nanoparticles and liposomes. Larger particles of zinc and titanium oxide used in topical skin-care products have been shown to be able to penetrate the stratum corneum barrier of rabbit skin with highest absorption occurring from water and oily vehicles. This could also apply to manufactured nanoparticles.

Can nanoparticles gain access to the epidermis after topical exposure, the first step in a toxicological reaction? Exposure to metallic nanoparticles, whose physical properties would allow them to catalyze a number of biomolecular interactions, potentially could produce adverse toxicological effects. More information is required regarding the efficiency of decontamination of nanoparticles from skin since solubilization and dilution, the two hallmarks of post-exposure decontamination, might be less efficacious for these solid structures.

Research should address the effects of dermal and systemic exposure to a number of types of nanoparticles in the skin. The skin is a primary route of potential exposure to toxicants, including novel nanoparticles. However, there is no information on whether particles are absorbed across the stratum corneum barrier or whether systemically administered particles can accumulate in dermal tissue. Nanoparticles may traverse through the stratum corneum layers at varying rates due to particle size or become sequestered within the epidermis to increase their exposure time to viable epidermal keratinocytes.

Nanomaterials are difficult to obtain in large quantities; therefore, it is best to conduct *in vitro*tests to estimate *in vivo* starting doses for toxicity testing. At least three or four concentrations with controls should be used in all *in vitro* systems. These data would provide a preliminary, but relevant, assessment of both systemic *exposure* after topical administration as well as cutaneous *hazard* after both topical and systemic exposure, two essential components of any risk assessment.

Cell Culture

Human epidermal keratinocyte (HEK) monolayers can be affected by nanoparticle interactions. It has already been shown that changes in biomarkers of viability and toxicity can occur with exposure to multi-wall carbon nanotubes.

Cytotoxicity endpoints should be evaluated:

- Cell viability-metabolic markers such as mitochondrial reduction of tetrazolium salts into insoluble dye (MTT),
- Decreased cell viability-membrane markers like neutral red uptake into cell lysosomes, trypan blue exclusion and cell attachment/cell detachment, and
- Pro-inflammatory cytokine affects measured by TNFα, IL-8, IL-6, IL-10, or IL-1β.

Genomics and proteomics assays could be used to explore the mechanism behind the toxicity. However, caution must be taken when using carbon black or any other material as a control because complications may occur.

Carbon can adsorb the viability dyes, such as neutral red, and interfere with the absorption spectra. False positives will occur. The type of carbon black used is extremely important. For instance, ultrafine carbon black has been utilized in inhalation studies but dosing in cell culture gives different results, especially when conducting viability and cytokine assays.

Three dimensional skin cell cultures are also available commercially. They have shown to be able to predict irritation but may significantly overestimate absorption or penetration.

Flow-through Diffusion Cell Studies

Diffusion cell system consists of flow-through diffusion blocks each containing multiple Teflon cells perfused by a constant temperature circulator through a Silastic oxygenator, an automatic fraction collector, and a desiccant. Circular fresh skin from pigs (pig skin mimics human skin and eliminates the extreme variability seen with random source human skin) or humans are placed epidermal–side up in Teflon flow-through diffusion cell. Compound containing nanoparticles is dosed on the epidermal side whilst the dermal side in each cell is bathed with receptor fluid at a set flow rate. The perfusate is collected at defined intervals up to 24 hrs and nanoparticles flux in the perfusate can be assessed by radioactivity counting, fluorescence, or UV detection. The skin surface can then be swabbed to remove non-absorbed surface particles and then tape stripped to remove a stratum corneum sample to assess nanoparticle penetration into this outermost epidermal layer. Serial sectioning of the skin can also be carried out.

Isolated Perfused Porcine Skin Flap (IPPSF)

The isolated perfused porcine skin flap (IPPSF) would be an ideal model to study the absorption and toxicity of nanomaterials. The IPPSF has an intact functional microcirculation, a viable epidermis and dermis and can be well controlled. A single-pedicle, axial pattern tubed skin flap is obtained from the abdomen of pigs following surgical creation of the flap perfused primarily by the caudal superficial epigastric artery and its associated paired venae commitantes. The IPPSF is transferred to the perfusion apparatus that is a custom designed temperature and humidity-regulated chamber. Nanomaterials can be topically dosed to the skin surface and perfusate samples collected over an eight hour period and assessed for nanoparticle flux.

Other acute toxicity *in vitro* assays are available but are used to test corrosives (rat transcutaneous electrical resistance (TER), commercially available EPISKIN, Epiderm and Corrositex) and irritation (EPISKIN, and Epiderm). However, the major traditional endpoint for skin toxicity is using the cell viability assay MTT reduction that has been shown to be unpredictable with nanomaterials due to marker interactions with nanoparticles.

Mucosa

Mucosa is the moist tissue that lines particular organs and body cavities throughout the body, including the nasal cavity, oral cavity, lungs, vagina and gastrointestinal tract. Potentially one of the most important portals of entry for nanoparticle exposure (excluding the nasal cavity and lung, which has been detailed above) is the gastrointestinal tract. Either accidental or intentional exposure via oral administration to the GI tract can lead to significant exposures. Efficient uptake of nanoparticles via the GI tract has been well documented in oral feeding studies and gavage studies using particles ranging from 10 nm to 500 nm. In these studies nanoparticles translocated through the mucosal lining and epithelial barrier of the intestine and were associated with the GALT (gastroinstetinal associated lymphatic tissue) and circulatory system within as little as 60 minutes time.

Intestinal epithelium can be studied using a variety of methods including immortalized cell-lines and tissue constructs. An example of immortalized cell-lines used to study the uptake of materials across the intestinal epithelial barrier include Caco cells, which have been used in many pharmaceutical studies to determine intestinal permeability.

These assays could be adapted for use in *in vitro* translocation rate studies or for developing a mechanistic understanding of the translocation process. IEC-6 and IEC-18 cell lines have been used extensively in mechanistic studies of the intestinal epithelial lining as well and may represent useful tools for nanoparticle research. These cells have been used to measure the activation of various signal pathways after toxicant exposure, as well as cytokine and ROS/RNS release.

Dependent upon the specific application, vaginal and oral-lining exposure may be possible although it is unlikely that these, in general, would represent significant portal of entry exposure routes. However, there are a variety of cell-lines and tissue constructs or models available for study of translocation and impact of nanoparticle exposure through these routes.

Cellular Assays

Study of target organs distal to the site of deposition presupposes that there is translocation and redistribution of nanoparticles away from the portals of entry in the lungs, skin or gut. As discussed above, potential target organs include blood, endothelium, neural tissue, heart, kidney, liver, and spleen.

Endothelium

The endothelium is represented by a thin layer of cells lining the vasculature throughout the body. It has been demonstrated that ultrafine particles or nanoparticles may have a wide range of effects on the endothelium. *In vitro* cultures may be useful in elucidating mechanistic information about transport across the alveolo-capillary or blood-brain barrier and on endothelial cell effects. Cultured endothelial cells are well suited to determining the effects nanoparticles may have on RNS production which has been demonstrated to play a significant role in the homeostasis of the vasculature.

Blood

In vitro studies using fractionated blood products (isolated red blood cells, platelets, leukocytes, or serum with complement) can be utilized in evaluating the effect on circulating blood. Activation of platelets, red blood cell interactions, production of ROS/RNS, cytokine/chemokine release from leukocytes, and complement activation are relevant endpoints to evaluate for nanoparticles. It has been demonstrated that nanoparticles have the ability to enter the circulatory system once translocation from site of entry has occurred.

Spleen

The spleen is a major site of immune processing and

lymphoid maturation, and accumulation of particles in the spleen may have consequences for immune responses and immunopathology. Spleen cells can be isolated and studied for the effects of nanoparticles *in vitro*. Endpoints could include antigen processing and immune responsivity *in vitro,* markers of lymphoid cell differentiation, and functional aspects such as dendritic cell function and lymphocyte proliferation.

Liver

The liver is a complex organ and is structurally and functionally heterogeneous. The liver is the major site for biotransformation and defence against foreign materials and xenobiotics. It is an integral structure having two separate blood supplies, many different cell types, and many different functions. Liver injury, due to nanomaterials, may be characterized based on histologic lesions, such as inflammation or necrosis. Injury to the liver may also be characterized at the molecular level. Some of the most common mechanisms of hepatocellular injury are via the cytochrome P450 metabolic pathways. The liver can excrete materials into the bile; therefore, the biliary system may be exposed as well. *In vivo* there are reports that a variety of different toxins cause hepatocellular injury by a range of different mechanisms such as cytochrome P450 activation, alcohol dehydrogenase activation, membrane lipid peroxidation, protein synthesis inhibition, disruption of calcium homeostasis, and activation of pro-apoptotic receptor enzymes. Every effort should be made to use human derived cells for *in vitro* assays, because these studies could be used to predict toxicity in humans. However, there is a considerable human variability in enzyme function.

Primary Human Hepatocytes

Primary human hepatocyte cultures are available commercially with well-characterized metabolic profiles and a full complement of metabolizing enzymes. Availability of human cells is limited due to the increase in demand for liver transplantations.

Isolated Perfused Liver

This complicated model system would be suitable for the study of nanomaterials, because it is the closest model that mimics *in vivo* and would allow for a detailed characterization of particle distribution within the organ.

Liver Slices

Modern techniques of precision slicing have allowed liver slices to become a good model, because it retains the normal tissue organization which may be particularly critical for nanoparticle studies.

Collagen Sandwich Cultures

In this model system, the structural and functional integrity is retained for several days. The bile canaliculi are well preserved, and release of the enzymes alanine aminotransferase and aspartate aminotransferase can be evaluated.

In general, using these *in vitro* systems, hepatic metabolism can be studied using isolated hepatocytes and cell lines and evaluated for changes in CYP450. Microsomes may be used in screening nanomaterials for metabolism using LC/MS to identify metabolite formation. Subcellular fractions, liver slices and whole liver homogenates may be used to evaluate liver function and toxicity. To study the effects of nanomaterials on hepatic function specific endpoints, such as enzyme systems, mitochondrial function, albumin synthesis, cell detachment, gene and/or protein expression, and membrane damage should be considered.

Mechanistically distinct endpoints could be utilized, such as cell morphology, viability, membrane damage, alamar blue metabolism, ATP content, covalent protein binding, peroxisomal proliferation, and GSH content. Other biomarker identifications, such as transcription and proteomic profiling should be studied. However, biomarkers may have limitations because immortalized cell lines are genotypically and phenotypically different from the organ itself. Also, hepatocyte cell cultures represent a single cell system and will only provide information on events that directly affect the cell itself. Some of these test systems may be able to predict the toxicity of nanomaterials as long as the assumptions and limitations are realised.

Nervous System

Central Nervous System

In vitro systems to study the effects of particles on the nervous system could include culture of neurons and addition of nanoparticles to determine effects on neuronal function. Endpoints could include ROS/RNS production, apoptosis, metabolic status, effects on the action potential and ion regulation in general.

Microglial cells are a type of macrophage found in the brain, and they may be involved in handling any nanoparticle that gets to the brain. The responses of microglial cells to nanoparticles should be studied along the lines of those described for macrophages in the lung section. Other cells that could be studied for effects of nanoparticles are astrocytes, glial cells that have a number of important roles that influence the behaviour of neurons, and oligodendrocytes which provide support to axons by producing the myelin sheath, which insulates the axons.

Peripheral Nervous System

The skin and the other portals of entry and target organs will have a nerve supply and the skin for example, has sensory nerves that are present near the surface of the body. Nanoparticles have the potential to gain access to these nerves and be transported or affect them in a number of ways. This could be studied by using neuron culture of peripheral nerves and studying the effect of nanoparticles for various relevant endpoints (*e.g.*, dorsal root ganglion neurons). In the autonomic nervous system both sympathetic and para-sympathetic neurons can be cultured, and effects of nanoparticles on their viability, metabolism, electrical activity and ionic homeostasis could be studied.

Heart

Cardiac function could be altered by nanoparticles that find their way into the heart muscle from the microcirculation. Cardiomyocytes can be cultured and effects on their general viability, ionic homeostasis and metabolism could be ascertained. Additionally, cardiomyocytes beat with regular rhythm *in vitro,* and the effects on this could be measured and any effects examined as to mechanism.

Kidney

The kidney is a major filtering system to eliminate toxicants from the bloodstream and it has been demonstrated that nanoparticles can be excreted via the kidney. Whether there are adverse effects of nanoparticles on the kidneys is unknown but this can be evaluated using *in vitro*techniques. Permeability assays used in pharmaceutical studies can be evaluated for use in measuring translocation and penetration through the renal tubules. Effects on the epithelial tubules and vasculature can be evaluated with existing cell culture techniques.

A variety of endpoints can be evaluated, including signal transduction response, oxidative stress, cellular viability, ion channel flux, modulation in the release of growth factors and proteinases as important indicators to renal homeostasis. Several models exist that may be useful to investigate nanoparticle-kidney interactions including renal tissue slices to evaluate translocation, oxidative stress, signal transduction responses and toxicity. Immortalized cell-lines are an inexpensive alternative to kidney slice models and may provide mechanistic information on the cytotoxicity of nanomaterials. Cells derived from isolated glomeruli, distal tubule/ collecting ducts, proximal tubule or proximal nephrons have been well characterized and are commercially available.

An example is the use of HEK-293 cells (human embryonic kidney cells) to evaluate cytotoxicity of chemicals. MDKK cells and LLC-PK1 cells have been used extensively in *in vitro* mechanistic studies and can be utilized to evaluate effects of nanoparticles.

Non-Cellular Assays

Durability

The ability of a particle to persist contributes to its ability to accumulate as dose. In fibre toxicology, there are well-documented dynamic and static protocols for assessing this property of durability *in vitro*, using Gambles balanced salt solution. These fibre protocols could be modified to allow measurement of durability of nanoparticles *in vitro*.

Complement Activation

The complement system is a protein cascade that has evolved to detect foreign, mostly microbial, surfaces. It is, however, activated by asbestos and by carbon nanoparticles. Their high surface per unit mass and surface activity may mean that other types of nanoparticles might be potent at activating the complement system. This might modify the response by opsonising the particles (C3b) or causing inflammation by the production of anaphylatoxin (C5a). Studies on the ability of nanoparticles to activate the complement system are therefore warranted.

Adsorptive Properties

The large surface area of nanoparticles means that they can adsorb proteins. Adsorption of different proteins might occur with different nanoparticle surfaces, and this could modify how they are handled by macrophages and other cells. This could therefore be a focus of study.

Free Radical Production

Most, probably all, pathogenic particles generate free radicals in cell-free systems, and this ability to cause oxidative stress contributes to their ability to initiate inflammation, and cause cell injury and genotoxicity. The free radicals can arise as a consequence of stable radicals at the particulate surface (quartz), redox cycling of ionic transition metal via the Fenton reaction (*e.g.*welding fume), or by unknown surface mechanisms (nanoparticle carbon black). The ability of particles to generate free radicals can be assessed by a range of assays, including plasmid DNA scission, electron paramagnetic spin resonance and 8-OH-dG production in 'naked' DNA or a DCFD assay for *in vitro* ROS production.

Computational Toxicology

In addition to establishing screening methods mentioned above, efforts to assess risk associated with engineered nanomaterials or other environmental stressors should include a collection of new technologies called computational toxicology. Computational toxicology is defined as the application of mathematical and computer models and molecular biology approaches to improve prioritization of data requirements and risk assessments for environmental protection.

This approach involves four areas:

1. Computational chemistry which refers to physical-chemical-mathematical modeling at the molecular level and includes topics such as quantum chemistry, force fields, molecular mechanics, molecular simulations, molecular modeling, molecular design, and cheminformatics;
2. Molecular biology which allows for the characterization of genetic constituency and the application of wide coverage technologies, such as genomics, proteomics, and metabonomics, to provide the key indicators of cellular and organismal response to stressor input;
3. Computational biology or bioinformatics, which involves the development of molecular biology databases and the analysis of the data;
4. Systems biology which refers to the application of mathematical modeling and reasoning to the understanding of biological systems and the explanation of biological phenomena.

Computational toxicology is designed to increase the capacity to prioritize, screen, and evaluate materials by enhancing the ability to predict their toxicity. In addition to the "omics," quantitative structure-activity relationships (QSARs) developed in physical organic chemistry should be evaluated as to whether they can aid in predicting the structure-property relationship of nanomaterials.

These multidisciplinary models could then be considered in a source-to-outcome continuum from environmental release through entire concentration, exposure concentration, target organelles, early biological effects, and adverse outcome. While finding the relationship between structure and toxicity of nanomaterials using computational toxicology is likely a long time away, it is well to keep these models in mind while developing screening methods and to use current methods to validate and inform their development.

RESEARCH GAPS

There is a paucity of data on the effects of nanoparticles on these different target cells and their respective endpoints *in vitro*. We therefore identify an urgent research need to obtain more information about nanoparticles in all of these systems.

We do, however, identify some pressing needs and these include:

- *In vitro* assays need to be used to determine important parameters that drive the toxicity and translocation potential of nanoparticles *e.g*. size, surface area, surface reactivity, etc.
- Decisions have to be made regarding the most appropriate and useful *in vitro* endpoints and their relative utility and importance (*e.g*. translocation, generation of ROS, cytokine release, cytotoxicity). A ranking of these *in vitro* assays in order of relevance and utility should be attempted.
- *In vitro* data should be used to develop a paradigm for nanoparticle toxicity that predicts the toxicity based on measurement of *in vitro* parameters; when mature, this paradigm could be critically tested *in vivo*.
- Toxicokinetic data should be used to select the target cells and systems that are appropriate and to select plausible dose levels to use in the *in vitro* assays.

- The effect of nanoparticle form (*e.g.* singlet particles or aggregates, use of surfactant) should be determined and the nanoparticle dose should be characterised as much as possible, *e.g.* regarding surface area, metals, etc.
- There is a pressing need to prepare and choose appropriate benchmark materials for *in vitro*testing.
- How do we interpret *in vitro* results without appropriate mechanistic information from *in vivo*models?

Recommendations

- *In vitro* tests are recommended as they provide a rapid and relatively inexpensive way to assess the potential toxicity of nanoparticles; there are, however, well-documented drawbacks of the *in vitro* assays such as their relative simplicity and the high doses commonly utilized.
- We recommend that "benchmark" particle controls be utilized in all studies such as crystalline silica and respirable TiO_2.
- Non-cellular tests including nanoparticle durability, complement activation, adsorption and free radical production can all yield valuable data on potential harmfulness of nanoparticles; computational toxicology may also make a contribution.
- Attention should be given to the potential confounding effect of adsorption of proteins or assay constituents onto the nanoparticles surface.
- Various cell-based systems are available with varying benefits and drawbacks, including single cell cultures of cell lines and freshly-derived cells, co-cultures, organ cultures (*e.g.* tracheal explants) and heart-lung preparations.
- The lung is a key target organ and so lung epithelial cells, macrophages, immune cells and fibroblasts represent key cells for nanoparticle effects with specific

regard to inflammation, immunopathology, fibrosis, genotoxicity, microbial defence and clearance.

- Skin represents a target for nanoparticles, especially from nanoparticles in cosmetics and a number of in vitro test systems are recommended including keratinocyte culture, Flow-through Diffusion Cell, and Isolated Perfused Porcine Skin Flap (IPPSF).
- Mucosa, the moist tissue that lines the nasal cavity, oral cavity, lungs, vagina and gastrointestinal tract also represents a potential target for nanoparticles and various *in vitro* systems are available for testing and should be utilized.
- The tendency of nanoparticles to gain access to the vasculature means that endothelium and components of the blood are potential targets of nanoparticles and these can be studied *in vitro*and we urge that this pathway receive special attention.
- The spleen, kidney, heart and liver will be target organs for bloodborne nanoparticles and we advise that a number of *in vitro* test systems are available to model effects in these organs.
- Transfer of nanoparticles to the brain and interactions with the autonomic nervous system in the lungs have been reported and we strongly recommend that *in vitro* models be utilized to study the impact of nanoparticles in these important neural cells.

Dosimetry

Since mass may not be the proper dose metric for comparing the toxicity of fine vs. ultrafine particles, characterization of the test material should also include surface area per mass and particle number per mass.

For practical purposes, dose could be monitored as mass delivered/animal or mass inhaled/animal and then be converted easily to a surface area or particle number dose as necessary, provided the correlation between these three particle parameters is available.

Benchmark Material

To place any pulmonary response to exposure to a given nanomaterial in perspective, results should be compared to those for particles of well-defined toxicity. Such benchmark materials could include nano-sized TiO_2, carbon black, or crystalline silica.

These benchmark materials should be characterized for surface area and particle number per mass, as well as for particle size and with respect to chemical purity and crystallinity to allow comparisons to be made using a variety of dose metrics.

Exposure Concentration

It is recommended that a minimum of three exposure levels be used. Information regarding the actual anticipated exposure levels in humans would be useful in determining the exposure concentration range to be evaluated. However, such information for nanoparticles is often lacking.

In all cases, similar exposure concentrations of the test and benchmark materials should be used, and the various dose metrics discussed above should be considered when choosing the exposures for the benchmark and test materials. It is recommended that the highest concentration chosen should exhibit toxicity with the benchmark material.

Exposure Duration

For intratracheal instillation or pharyngeal aspiration, a single exposure to the nanomaterial is sufficient for Tier 1 studies.

Caveat: consider high dose and bolus effect! For inhalation, a two week exposure is recommended, although shorter exposures, perhaps at higher concentrations, should be done if this mimics human exposures.

Pulmonary Parameters

Pulmonary responses should be monitored 24 hours to 28 days post-exposure. A suggested time course could be 24 hrs, 1 week and 28 days post-exposure.

Pulmonary Endpoints

The degree/intensity and duration of pulmonary inflammation and cytotoxic effects following nanoparticle exposures are important endpoints for assessing the toxicity of a test nanoparticle.

- *Bronchoalveolar lavage (BAL) damage markers*: BAL profile. This method samples the cells and fluid from the bronchoalveolar space and allows the assessment of inflammation by quantification of cell numbers and types and components of the fluid phase. In addition, considerable extra information can be gained by various *ex vivo* manipulations of the BAL cells, *e.g.*, gene expression, phagocytic potential, etc. Other BAL damage markers include BAL lactate dehydrogenase levels (as a measure of cytotoxicity), BAL protein levels (increases in BAL fluid protein concentrations generally are consistent with enhanced permeability of vascular proteins into the alveolar regions, indicating a breakdown in the integrity of the alveolar-capillary barrier), and BAL alkaline phosphatase levels (as a measure of Type 2 alveolar epithelial cell toxicity). Methodologies for cell counts, differentials, and pulmonary biomarkers in lavaged fluids have previously been described.
- *Oxidative stress markers*: ROS/RNS. Reactive oxygen and nitrogen species have been implicated in DNA damage and induction of inflammatory cytokines and growth factors. Acellular BAL fluid levels of gluthathione, total antioxidants, or nitrate/nitrite (a measure of nitric oxide production), lipid peroxidation of lung tissue, or *ex vivo* measurement of ROS/RNS from BAL cells can be employed to monitor oxidant generation and oxidant stress. Methodologies for oxidative stress markers have been described.
- *Histopathology*: Description of the general effects of treatments on the lungs should include endpoints such as presence of dust-laden macrophages, cellular

infiltrates and hyperplastic changes in the epithelium. It is recommended that the entire respiratory tract be evaluated for adverse pathological effects. This would include the upper respiratory tract–the nose, larynx and upper airways; the lower respiratory tract and lymph nodes; and the pleural region. Histopathological observations in a Tier 1 process would focus primarily on inflammatory responses and the development of fibrosis. Fibrosis can be determined in lung tissue by specific staining of collagen in histopathological slides, or by qualitative and quantitative histopathology.

- *Cell proliferation*: Increased cell division plays a key role in pathological responses and can be determined in epithelial or mesothelial cells by uptake of labeled nucleotide precursors, such as tritiated thymidine or BrdU. Recommended experiments are designed to measure the effects of particle exposures on airway and lung parenchymal cell turnover in rats following exposures. Groups of particulate-exposed rats and corresponding controls can either be pulsed or implanted subcutaneously with minipumps containing 5-bromo-2'deoxyuridine (BrdU) dissolved in a sodium bicarbonate buffer solution. Methodologies for cell proliferation studies have previously been described. An alternative method to BrDU staining is the PCNA staining method. Proliferating cell nuclear antigen (PCNA) is a nuclear protein associated with cell proliferation and has been used to discriminate via immunohistochemistry proliferating cells in numerous tumor types including those found in the lung.

Intratracheal Instillation or Pharyngeal/Laryngeal Aspiration Studies

As discussed above, inhalation is the most physiologically relevant and therefore preferred method of pulmonary exposure for hazard identification and to obtain dose-response data.

However, both intratracheal instillation of nanomaterials suspended in an appropriate vehicle and pharyngeal or laryngeal aspiration (with appropriate caveats) are considered to be acceptable methods for pulmonary exposure to evaluate the relative toxicity of the test material. Similar to studies using aerosol exposures, the following pulmonary endpoints should be evaluated in a Tier 1 testing strategy approach for assessing lung hazards to nanoparticles:

- BAL damage markers
- Oxidative stress markers
- Histopathology
- Cell proliferation

Other Organ Endpoints

Exposure to nanoparticles *via* the respiratory tract includes a high probability of translocation to other organs and tissues–depending on nanoparticle size and surface chemistry. Although translocation rates may be very low, localization at sensitive sub-cellular sites (*e.g.*, mitochondria) could result in adverse responses directly induced by the nanoparticle. Alternatively, or in addition, potential oxidative stress and inflammatory responses elicited by nanoparticles in the respiratory tract may result in the release of mediators which can lead to indirect secondary effects in extra-pulmonary organ systems. Thus, it is essential to include an evaluation of potential effects in remote organs and tissues, such as liver, spleen, bone marrow, heart, kidney, and CNS, in the Tier 1 evaluation.

Histopathological examination of extra-pulmonary tissues should be mandatory; however, this alone may only show significant effects following longer-term or very high exposures. Therefore, consideration should also be given to determining organ-specific endpoints, such as acute phase proteins and coagulation factors, for effects on the cardiovascular system, immune response assays for effects on the spleen, and immunohistochemical staining for dopaminergic neurons in brain sections to evaluate neurogenic effects. Additional functional assays (*e.g.*, measurement of heart rate variability) may be considered, but since they require specific equipment and expertise, these are not mandatory for Tier 1 studies.

Pulmonary Exposure–Tier 2

Research results showing that nanoparticles can translocate from the portal of entry, the respiratory tract, via different pathways to other organs/tissues makes them uniquely different from larger-sized particles in that they may induce direct adverse responses in remote organs.

In particular, such responses may be initiated through the interaction of nanoparticles with sub-cellular structures following endocytosis by different target cells. For this reason, special attention needs to be given to recognizing such effects, which in the healthy organism are probably very subtle initially, even not detectable, but could have serious consequences in a compromised organism or a compromised organ. Examples are effects of anthropogenic ultrafine particles in asthmatics, in people with cardiovascular diseases, in the elderly and very young.

Complementary Tier 2 studies can be used to obtain more data for hazard identification as an initial step for risk assessment. Tier 2 studies will provide additional information to either characterize further effects seen in Tier 1 studies or to obtain new data using specific models of susceptibility. Ideally, studies should be performed using inhalation exposures as a first choice, in particular if a positive response was seen in Tier 1 studies when intratracheal instillation or pharyngeal/laryngeal aspiration was used. If insufficient amounts of material are available, multiple low dose (1–10 μg/kg body weight) exposures of the respiratory tract using the above-mentioned non-inhalation methods can be applied, *i.e.*, dosing once or twice/week for 4 weeks with 2–3 months follow-up.

Animal Models

The absence of an effect of nanoparticles in a normal animal model does not imply that there will be no effect in a model which exhibits enhanced susceptibility. Increased susceptibility can be due to a number of factors, including age, disease, altered organ function, genetic polymorphism. Respective animal models include exposures of senescent, transgenic and knockout animals and animals with compromised organ systems (*e.g.*, hypertension; diabetes models; immuno-compromised, infectivity models).

In general, susceptibility models include compromised functions of the respiratory tract, CNS, cardiovascular system (dysfunction of endothelial cells, platelets), bone marrow, and/or kidney. It is essential that the models used are relevant to the human disease state and that a respective animal model has been validated in the peer-reviewed literature. In addition to obtaining information for identifying nanoparticle hazards, Tier 2 studies will also provide data on underlying mechanisms which can be used in concert with the mechanistic *in vitro* studies.

Multiple Exposures

As indicated above and under Tier 1 studies, repeated inhalation exposures are the first choice for realistic dosing. Information about the physicochemical nature of airborne nanoparticles at the workplace, or anticipated exposure of the general public (consumer), is essential to mimic the same for animal exposures.

Issues include: are the nanoparticles aggregated or singlets; what is the diameter in the airborne state; what is known about other chemical characteristics? Generation and monitoring of airborne nanoparticles for inhalation exposures requires special equipment and expertise. As an alternative to inhalation, intratracheal instillation, oropharyngeal or laryngeal aspiration can be used. However, unless the nanoparticles are coated and made "soluble" in physiological solutions, artifacts due to aggregation may occur when using these non-inhalation methods. In addition, the upper respiratory tract is circumvented by the non-inhalation methods, thereby eliminating potential neuronal nanoparticle translocation to the CNS.

To address this concern, one could expose the animal to nanoparticles by nasal instillation. Techniques for tracking the uptake of nanoparticles by sensory neurons would be similar to those discussed in the section concerning deposition, translocation, and biopersistence. Furthermore, the impact of coating for purposes of "solubilizing" the nanoparticles has to be carefully considered in cases where anticipated human exposure is to the uncoated material, since cellular uptake, translocation and effects will be affected by the surface coating.

Daily repeated inhalation exposures over four weeks are suggested for the Tier 2 studies, with an up to 3-month post-exposure observation period, including interim post-exposure sacrifice days. With respect to the non-inhalation methods of exposure, dosing can occur 1–2 times/week over a 4-week period, followed by the post-exposure observation period. Special attention needs to be given to the selection of exposure concentrations (inhalation) and doses (instillation, aspiration). Knowledge about anticipated human exposure will be extremely valuable, and use of predictive particle deposition models (*e.g.*, MPPD model) can be used to determine realistic exposure concentrations/doses.

The Tier 2 studies are aimed at obtaining additional information with respect to the biokinetics of the nanomaterials following exposure of the respiratory tract the stability of nanoparticles in the organ system (*e.g.,in vivo* change of surface chemistry, bioavailability of core material), and the potential acute and sub-acute effects in the mammalian organism, including genomic and proteomic evaluation. Other endpoints are the same as listed under Tier 1 studies for the respiratory tract.

Deposition, Translocation and Biopersistence Studies

Exposure to airborne nanoparticles *via* the inhalation route leads to deposition in the various compartments of the respiratory tract according to probabilities dependent on three important parameter groups: aerodynamic and thermodynamic nanoparticle properties, breathing pattern, and the three-dimensional geometry and structure of the respiratory tract. Deposition probability of nanoparticles below a thermodynamic diameter of 500 nm increases with decreasing size because of the increasing diffusion velocity leading to an increased deposition in the small airways and the alveoli, in particular. Below 20 nm, the location of deposition of nanoparticles changes to the upper respiratory tract because of their even higher diffusion velocity. Currently existing computer codes provide a first estimate of deposition probabilities in the various regions of the respiratory tract which may require modification if there are indications that nanoparticles may undergo changes of their aero- and/or thermodynamic properties not being considered in those codes.

Once deposited, insoluble nanoparticles undergo clearance mechanisms specific to the region of the respiratory tract; *i.e.*, at all regions nanoparticles will interact with proteins of the epithelial lining fluid potentially forming complexes which are likely to affect their subsequent metabolic fate and biokinetics.

On the epithelium of conducting airways, mucociliary clearance provides a rapid transport, to the larynx for further transport into the gastro-intestinal tract and excretion. Note, however, that there also occurs long-term retention in human airways for a fraction of nanoparticles which increases with the decreasing size of nanoparticles giving rise to cellular uptake. On the epithelium of the alveolar region there is no rapid transport so that phagocytosis by free phagocytes can occur with subsequent slow clearance to the larynx, but because of the limited capability of macrophages to recognize nanoparticles, endocytotic processes and trans-cellular transport by other cells like epithelial type I + II cells, become prominent together with para-cellular transport mechanisms across tight junctions under inflamed conditions. These mechanisms result in translocation of nanoparticles into the interstitium, lymphatic drainage and possible translocation across endothelial cells into capillaries.

This access to the blood circulation provides accumulation and possible adverse reactions in secondary target organs such as the cardiovascular system, liver, spleen, bone marrow, central nervous system, endocrine organs and interaction with endothelial cells, platelets and immuno-competent cells in the circulation.

Besides the direct effects of translocated nanoparticles on secondary organs, indirect effects may occur as well triggered by interactions of nanoparticles at their site of retention in the respiratory tract with adjacent biological systems like cells, fluids, proteins and extracellular matrix. Subsequent cell activation can lead to release of cytokines and other mediators which subsequently diffuse into the circulation to induce adverse responses in secondary target organs. Because the underlying mechanisms of nanoparticle translocation and accumulation or mediator response in secondary target organs are not fully understood, the determination of nanoparticle kinetics should be a high priority.

The scenario described above relates to biopersistent nanoparticles which maintain their particulate state; however, nanoparticles even when they may be insoluble in water may not persist as a solid particle in cells and body fluids but may fully or partially (*e.g.*, surface coating) disintegrate/dissolve, so that eventually a biopersistent core with different particle properties/toxicities is retained. Because of the high diversity of emerging nanoparticles, studies of their biopersistence should be regarded of high priority.

Methodologies for such biopersistence studies include:

- *Radio-labeled or fluorescently or magnetically tagged monitoring in lungs and various organs:* Radio-labeling of nanoparticles specifically with a radio-isotope of one of the contextual chemical elements of the particle matrix is the gold standard for studying translocation kinetics allowing for time-efficient, extremely high sensitivity and specificity measurements, particularly when they aim to account for a balance of the entire distribution in the body and in excretions. Also fluorescent or magnetic labeling provides a powerful means of high sensitivity and specificity to determine translocated fractions in various target organs. Extreme care should be taken; however, to assure that any label stays firmly with the nanoparticles, otherwise the results will be severely flawed. Also, evidence is required that the process of labeling does not modify the nanoparticle in its function and its surface since surface is the predominant interacting substrate with biological systems.

Labeling of nanoparticles will not be possible in every case. In this event tracking of nanoparticles by electron microscopy may be a suitable alternative to monitor the fate of electron-dense nanoparticles in the organism. The method of choice for evaluation is a quantitative morphometric approach, which is by far preferable over qualitative spotting images, which may lead to a wrong interpretation. In addition, it is strongly recommended to search for adequate alternate labeling techniques, which may not be obvious at the first glance but would still provide a feasible option, for example use of confocal microscopy with fluorescently labeled nanoparticles.

If the matrix of the nanoparticle provides a characteristic chemical element or compound, chemical analysis of this characteristic provides another strong alternative to track the fate of nanoparticles. Importantly, contamination and possible endogenous background levels require careful distinction as well as the determination of the lower level of sensitivity of the analytical approach.

Genomics and Proteomics

Nanomaterials distributing to different tissues of the body following deposition in the respiratory tract can potentially affect multiple cellular functions, and it will be difficult to determine with conventional assays what changes and adverse effects may have occurred. Use of genomic and proteomic analyses should be considered, the former providing information about specific mechanisms at the molecular level (*e.g.*, oxidative stress) and the latter linking this to the expression of proteins resulting in effects at the cellular and tissue level.

Results from such analyses are needed to help in the interpretation of responses.

Analysis of the results of these assays requires the input of bioinformatics which will help in the interpretation of elicited responses. Together, genomic and proteomic studies represent an effective strategy combining hypothesis-forming and hypothesis-driven research which is needed in the assessment of nanoparticle risks within the framework of a multidisciplinary team approach.

Effects on the Reproductive System, Placenta, and Fetus

The studies to assess Tier 2 reproductive effects following pulmonary exposures to nanoparticles should follow protocols similar to the OECD Guideline 422 for Testing of Chemicals. The test substance should be administered in gradual doses to several groups of male and female rats. Males should be dosed for a minimum of 4 weeks (which includes a minimum of 2 weeks prior to mating during the mating period and approximately 2 weeks post mating). Given the limited pre-mating dosing period in males, fertility may not be a particularly sensitive indicator of testicular toxicity and should be concomitant with a detailed histopathological analysis of the male gonads to assess impact on fertility and spermatogenesis.

Females should be dosed throughout the study–including 2 weeks prior to mating (with the objective of covering a minimum of 2 estrus cycles), the variable time to conception, the duration of pregnancy, and a minimum of 4 days after delivery, up to and including the day before the scheduled sacrifice. The duration of gestation should be recorded and is calculated from day 0 of pregnancy. Each litter should be examined as soon as possible after delivery to establish the number and sex of pups, stillbirths, live births, runts (pups that are significantly smaller than corresponding control pups), and the presence of gross abnormalities.

Live pups should be counted and sexed and litters weighed within 24 hours of parturition (day 0 or 1 post-partum) and on day 4 post-partum. In addition to the observations on parent animals, any abnormal behaviour of the offspring should be recorded.

Oral Exposure–Tier 1

It is possible that during the life of a nanomaterial (production, application, disposal, etc) it may appear in the water supply or be inadvertently ingested. If this is a concern, the effects of oral exposure to the nanomaterial should be investigated. Exposure should be by a single gavage at a dose which would represent the worse case human exposure. As with pulmonary exposure, for oral exposure the physical and chemical properties of the test material should be characterized in the form delivered to the test animal. Rats or mice are the recommended model system.

There is no preference concerning gender for such studies. The feces should be collected for four days post-exposure, and the amount of nanomaterial eliminated vs. retained should be determined. Particularly GALT, mesenteric lymph nodes and liver should be analysed for the presence of nanoparticles. If absorption of the nanomaterial from the gastrointestinal tract is near zero, then the systemic effects of oral exposure to that nanomaterial need not be evaluated. However, if significant absorption of the nanomaterial is evident, evaluation of systemic toxicity is recommended using histology and functional assays as described for the various organ systems after pulmonary exposure.

Injection–Tier 1

Some nanomaterials are being evaluated as drug delivery systems. In such a case, the potential toxicity of this nanomaterial after injection should be evaluated. Rats or mice are the recommended model system. There is no preference concerning gender for these studies. If possible, a tagged nanoparticle should be injected and its distribution to various organs (liver, spleen, heart, bone marrow, kidney, and lung) and elimination in the feces should be monitored for a week post-exposure.

Histology and functional assays (*e.g.*, mitochondrial function) of the various organ systems should be implemented as described following pulmonary exposure.

Skin Exposure–Tier 1

For skin absorption of nanomaterials, the most appropriate animal model should be used. Rats are most common but rabbits, guinea pigs and pigs are also used to assess toxicity and irritation. The rat and pig are recommended as the animal of choice. The rat being small and already having an established database in the field of toxicology by other routes of exposure could be used because the amount of nanomaterials needed would be less for this small species. Frequently, the domestic pig is utilized in absorption studies because the skin is anatomically, physiologically and biochemically similar to that of humans.

Twenty-four hours prior to dosing with nanomaterials, the area (10% of the body surface) on the back should be clipped to remove hair. Three doses at log intervals (mg/cm^2) plus controls (vehicle, no material controls and a positive control) should be applied to both normal skin and abraded skin to mimic how humans are exposed. The material should be applied in an occluded fashion because nanomaterials, unlike many chemicals, will not be absorbed into the skin immediately.

The occlusion device (site protection) used to prevent the nanomaterial from falling off should be attached to the surface of the skin by non-irritating tape. At least 4–6 animals per group/ dose plus controls and vehicle controls should be utilized over the duration of 24 hrs. At 0.5, 1, 2, 4, 8, and 24 hrs, the treatment sites should be scored for erythema and edema using the Draize test scores.

Skin biopsies should be taken for transmission electron microscopy that would identify cellular changes as well as localize the penetrated particles within the skin. Light microscopy could be utilized to assess the morphological alterations that could occur due to the acute toxicity of the nanomaterials but this will not detect nanomaterial localization.

For repeated exposures, nanomaterials should be applied daily for 5 or 7 days and could continue until 28 days. This could depend on the type and the amount of nanomaterials that are available. If a 28 day study is planned, then daily clinical observations should be conducted. At termination, hematology, clinical chemistry, evaluation of local lymph nodes and an immunotoxicology battery of tests should be performed. Standard full necropsy exam (liver, kidney, etc) should also be conducted. The specific goal of the study, dermal absorption or irritation, will dictate the specific study design (*e.g.* duration of study, samples collected).

Research Gaps

Significant research gaps exist; the first four included in the following list pertain to general informational needs on production, use, and exposure to nanomaterials that would be helpful in the design of toxicity test.

- What is being made and in what quantities in the nanotechnology industry?
- What exposure levels are likely in the workplace?
- What is/are the likely route(s) of exposure?
- What are occupational vs. environmental exposures?
- Radio-labeled particles are needed for investigation of deposition, translocation, and biopersistence. This requires specific lab that can work with and detect labeled materials; labeling not feasible for many materials.
- A source of reference nanomaterials should be available to researchers.

Recommendations

- Studies involving *in vitro* (non-cellular and cellular), pulmonary, oral, injection and dermal exposure are recommended in Tier 1 testing.
- It is essential that exposures in animal models be relevant to human exposures, when known, for production, use, and disposal.

- It is recommended that exposure route be relevant to anticipated human exposures for production, use, and disposal.
- It is recommended that the test material be fully characterized, preferably as delivered to the animal.
- Following pulmonary exposure, recommended endpoints to be measured include organ-specific markers of inflammation, oxidant stress and cell proliferation (*e.g.*, mitochondrial) and histopathology in the lung as well as measurement of damage to non-pulmonary organs.
- *Studies involving*:
  - Use of susceptible models,
  - Multiple exposures,
  - Evaluation of deposition, translocation and biopersistence,
  - Reproductive effects, and
  - Mechanistic genomic and proteomic techniques may be considered for Tier 2 testing.

CONCLUSION

Engineered nanomaterials presenting a potential risk to human health include those capable of entering the body and exhibiting a biological activity that is associated with their nanostructure. Nanomaterial-based products such as nanocomposites, surface coatings and electronic circuits are unlikely to present a direct risk as exposure potential will be low to negligible. Nanomaterials that are most likely to present a health risk are nanoparticles, agglomerates of nanoparticles, and particles of nanostructured material (where the nanostructure determines behaviour). In each of these cases, exposure potential exists for materials in air and in liquid suspensions or slurries.

Recognizing the early stage of understanding of the potential toxicity of nanomaterials and that little knowledge exists regarding specific nanomaterial characteristics which may be indicators of toxicity, the elements of a screening strategy outlined in this document include a significant research component. The range and extent of recommended testing reflect this developing state of knowledge. The elements of a screening strategy are clear, but the detailed approach will evolve and become more focused and selective as the results of these early-stage screening/research studies become available. A more thorough discussion of the 'elements' presented and the development of a more robust and detailed strategy will only be possible as knowledge increases. Elements of a nanotoxicity testing strategy which have been detailed in previous sections are summarized below.

PHYSICOCHEMICAL CHARACTERIZATION

Appropriate physicochemical characterization of nanomaterials used in toxicity screening tests is essential, if data are to be interpreted in relation to the material properties, inter-comparisons between different studies carried out, and conclusions drawn regarding hazard. The dependence of nanomaterial behaviour on physical and chemical properties places stringent requirements on physicochemical characterization and includes assessing a range of properties, including particle size distribution, agglomeration state, shape, crystal structure, chemical composition, surface area, surface chemistry, surface charge and porosity. Precise requirements will differ for *in vivo*and *in vitro* studies, and according to the material delivery route or method. In addition, characterizing human exposures introduces a third set of requirements.

A wide range of analytical methods are available that are applicable to nanomaterials, and multidisciplinary collaborations are encouraged to ensure appropriate methods are adopted. Particular consideration should be given to the use of Transmission Electron Microscopy, which in many cases can be considered the gold standard of nanoparticle characterization. In addition, information on nanomaterial production, preparation, storage, heterogeneity and agglomeration state should be recorded in all cases. Characterization of nanomaterials after administration *in vitro* or *in vivo* is considered the ideal in screening studies, although it currently presents significant analytical challenges. Characterization of the material as administered is therefore recommended for most screening tests.

Characterization of the nanomaterial solely as produced or supplied is only considered appropriate where the previous two approaches are not viable. In all screening studies, dose should be evaluated against appropriate metrics. The three principal physical metrics of interest are mass, surface area and number concentration of particles: given current uncertainty over the relevance of each, it is important that all three are measured or derivable in any given study.

In Vitro Testing Methods

In vitro tests of toxicity yield data rapidly and can provide important insights and confirmations of the mechanism of *in vivo* effects. We recommend that a wide range of *in vitro* tests be applied to the key research questions relating to the potential hazard associated with nanoparticles exposure.

A wide range of *in vitro* approaches exist that can be matched to specific questions relating to different aspects of nanoparticles toxicity. Non-cellular tests can provide information on aspects such as biopersistence, free radical generation by particle surfaces and activation of humoral systems such as the complement system; computational toxicology methods may also be useful. Cell-based systems can comprise cell lines and freshly derived primary cells in monocultures or co-cultures.

Organ cultures and heart/lung preparations are also potentially useful for studying nanoparticles effects and translocation. We recommend that the *in vitro* tests should reflect the different portals of entry and target organs that nanoparticles could impact and these include lung, skin, mucosal membranes, endothelium, blood, spleen, liver, nervous system, and heart. As always, care should be taken in interpreting data obtained from *in vitro* systems because of the high doses normally used *in vitro* and the impact of a bolus effect. There should be inclusion of appropriate benchmark particles to contextualize the results of *in vitro* assays. We recommend vigilance for artifactual effects peculiar to nanoparticles caused by their large adsorptive surface which can deplete cell products or assay constituents and thereby confound assay results. In addition to the utilization of existing test systems we suggest that new assays may be developed, for example to study transit of nanoparticles across cell layers.

In Vivo Testing Methods

For *in vivo* testing of nanomaterials, two tiers of studies are discussed. Tier 1 studies would involve pulmonary, oral, injection, and dermal exposure as would be relevant to the human exposure(s) of concern.

A critical initial step in *in vivo* testing is full characterization of the test material. Endpoints of concern for pulmonary exposure involve organic-specific markers of inflammation, oxidant stress, and cell proliferation and histopathology in the lung as well as measurement of damage to non-pulmonary organs. Tier 2 pulmonary exposure studies are recommended but not mandated.

These studies would provide useful information for a complete risk assessment of a nanomaterial.

Tier 2 studies include:

- Use of susceptible models,
- Effects of multiple exposures,
- Deposition, translocation and biopersistence studies,
- Evaluation of reproductive effects, and
- Mechanistic studies employing genomic and proteomic techniques.

The testing strategy for *in vivo* studies would be of greatest value for hazard identification and risk assessment if exposure dose, route of exposure, and particle characteristics closely modeled those of human exposure. Therefore, an understanding of the life cycle of a given nanomaterial, *i.e.*, exposure during production, upon use, and environmentally, is a critical research need.

3

Natural Colloids and Nanoparticles in Aquatic and Terrestrial Environments

MAJOR TYPES OF ENVIRONMENTAL COLLOIDS

The Scottish chemist Thomas Graham discovered that certain substances (*e.g.*, glue, gelatin, or starch) could be separated from certain other substances (*e.g.*, sugar or salt) by dialysis. He gave the name *colloid* to substances that do not diffuse through a semipermeable membrane (*e.g.*, parchment or cellophane) and the name *crystalloid* to those which do diffuse and which are therefore in true solution. Colloidal particles are larger than molecules but too small to be observed directly with a microscope; however, their shape and size can be determined by electron microscopy. In a true solution the particles of dissolved substance are of molecular size and are thus smaller than colloidal particles; in a coarse mixture (*e.g.*, a suspension) the particles are much larger than colloidal particles. Although there are no precise boundaries of size between the particles in mixtures, colloids, or solutions, colloidal particles are usually on the order of 10^{-7} to 10^{-5} cm in size.

CLASSIFICATION OF COLLOIDS

One way of classifying colloids is to group them according

to the phase (solid, liquid, or gas) of the dispersed substance and of the medium of dispersion. A gas may be dispersed in a liquid to form a foam (*e.g.*, shaving lather or beaten egg white) or in a solid to form a solid foam (*e.g.*, styrofoam or marshmallow). A liquid may be dispersed in a gas to form an aerosol (*e.g.*, fog or aerosol spray), in another liquid to form an emulsion (*e.g.*, homogenized milk or mayonnaise), or in a solid to form a gel (*e.g.*, jellies or cheese). A solid may be dispersed in a gas to form a solid aerosol (*e.g.*, dust or smoke in air), in a liquid to form a sol (*e.g.*, ink or muddy water), or in a solid to form a solid sol (*e.g.*, certain alloys).

A further distinction is often made in the case of a dispersed solid. In some cases (*e.g.*, a dispersion of sulfur in water) the colloidal particles have the same internal structure as a bulk of the solid. In other cases (*e.g.*, a dispersion of soap in water) the particles are an aggregate of small molecules and do not correspond to any particular solid structure. In still other cases (*e.g.*, a dispersion of a protein in water) the particles are actually very large single molecules. A different distinction, usually made when the dispersing medium is a liquid, is between lyophilic and lyophobic systems. The particles in a lyophilic system have a great affinity for the solvent, and are readily solvated (combined, chemically or physically, with the solvent) and dispersed, even at high concentrations. In a lyophobic system the particles resist solvation and dispersion in the solvent, and the concentration of particles is usually relatively low.

FORMATION OF COLLOIDS

There are two basic methods of forming a colloid: reduction of larger particles to colloidal size, and condensation of smaller particles (*e.g.*, molecules) into colloidal particles. Some substances (*e.g.*, gelatin or glue) are easily dispersed (in the proper solvent) to form a colloid; this spontaneous dispersion is called peptization. A metal can be dispersed by evaporating it in an electric arc; if the electrodes are immersed in water, colloidal particles of the metal form as the metal vapour cools.

A solid (*e.g.*, paint pigment) can be reduced to colloidal particles in a colloid mill, a mechanical device that uses a shearing force to break apart the larger particles. An emulsion is often prepared by homogenization, usually with the addition of an emulsifying agent. The above methods involve breaking down a larger substance into colloidal particles. Condensation of smaller particles to form a colloid usually involves chemical reactions-typically displacement, hydrolysis, or oxidation and reduction.

PROPERTIES OF COLLOIDS

One property of colloid systems that distinguishes them from true solutions is that colloidal particles scatter light. If a beam of light, such as that from a flashlight, passes through a colloid, the light is reflected (scattered) by the colloidal particles and the path of the light can therefore be observed. When a beam of light passes through a true solution (*e.g.*, salt in water) there is so little scattering of the light that the path of the light cannot be seen and the small amount of scattered light cannot be detected except by very sensitive instruments.

The scattering of light by colloids, known as the Tyndall effect, was first explained by the British physicist John Tyndall. When an ultramicroscope is used to examine a colloid, the colloidal particles appear as tiny points of light in constant motion; this motion, called Brownian movement, helps keep the particles in suspension. Absorption is another characteristic of colloids, since the finely divided colloidal particles have a large surface area exposed. The presence of colloidal particles has little effect on the colligative properties (boiling point, freezing point, etc.) of a solution.

The particles of a colloid selectively absorb ions and acquire an electric charge. All of the particles of a given colloid take on the same charge (either positive or negative) and thus are repelled by one another. If an electric potential is applied to a colloid, the charged colloidal particles move towards the oppositely charged electrode; this migration is called electrophoresis. If the charge on the particles is neutralized, they may precipitate out of the suspension. A colloid may be precipitated by adding another colloid with oppositely charged particles; the particles are attracted to one another, coagulate, and precipitate out. Addition of soluble ions may precipitate a colloid; the ions in seawater precipitate the colloidal silt dispersed in river water, forming a delta. A method developed by F. G. Cottrell reduces air pollution by removing colloidal particles (*e.g.*, smoke, dust, and fly ash) from exhaust gases with electric precipitators. Particles in a lyophobic system are readily coagulated and precipitated, and the system cannot easily be restored to its colloidal state. A lyophilic colloid does not readily precipitate and can usually be restored by the addition of solvent.

Thixotropy is a property exhibited by certain gels (semisolid, jellylike colloids). A thixotropic gel appears to be solid and maintains a shape of its own until it is subjected to a shearing (lateral) force or some other disturbance, such as shaking. It then acts as a sol (a semifluid colloid) and flows freely. Thixotropic behaviour is reversible, and when allowed to stand undisturbed the sol slowly reverts to a gel. Common thixotropic gels include oil well drilling mud, certain paints and printing inks, and certain clays. Quick clay, which is thixotropic, has caused landslides in parts of Scandinavia and Canada.

INTRINSIC PROPERTIES OF ENVIRONMENTAL COLLOIDAL PARTICLES

The nasal passages of mammals play an important role in reducing water loss through this pathway. Respiratory surfaces are a major avenue for water loss in air-breathing animals. The internalization of the respiratory surfaces in a body cavity such as the lungs reduces evaporative loss in terrestrial vertebrates. Because the body temperature of birds and mammals is generally higher than external temperatures, evaporative loss of water is greater. Warm expired air contains more water than the cooler inspired air, as the water holding capacity of air increases with temperature.

A mechanism, termed a temporal countercurrent system retains most of the respiratory water vapour by condensing it on cooled nasal passages during expiration. The nasal passages of mammals plays an important role in reducing the loss of water and heat from the body. The importance of the nasal passages in cooling expired air can be detected easily by placing your hand in front of your nose when breathing, and comparing this to putting your hand in front of your mouth when breathing.

In most animals, the majority of cells are not in direct contact with the external environment but are bathed by an internal body fluid. Homeostatic mechanisms hamper changes in an animal's body fluid, which both gives protection from harmful external environments and impedes quick exchange between intracellular compartments. The cells of the animal cannot survive much additional water gain or loss. Water continuously enters and leaves an animal cell across the plasma membrane, however, uptake and loss must balance. Animal cells swell and burst if there is a net uptake of water or shrivel and die it there is a net loss of water.

Other problems associated with osmoregulation are body and environment temperatures. The enzyme activity in the body function between temperatures of 0°–40° Celsius. The way animals deal with temperature and regulating it is by way of water loss. So animals in hot environments need to limit the amount of water loss due to evaporation and respiration. The importance of water in temperature regulating leads to conflicts and compromises between physiological adaptations to environmental temperatures and osmotic stresses in terrestrial animals.

Organisms in different environments utilize different structures in osmoregulation and excretion. The major structures involved are the integument, the respiratory surface, the kidney, and the salt gland. All animals use at least one of these structures in their osmoregulatory processes. The common characteristic in structures such as gills, skin, kidneys and the integument are cells called transport epithelia, anatomically and functionally polarized cells which determine the osmoregulatory capabilities of the structure, through properties such as permeability to various solutes. The integument functions in osmoregulation by acting as a barrier between the extracellular compartment and the environment to regulate water gain and loss, as well as solute flux. The permeablity of the integument to water and solutes varies from animal to animal.

Respiratory surfaces such as the alveoli of the lung, and gills in aquatic animals also serve in osmoregulation and excretion. Respiratory surfaces are the chief avenues for the excretion of carbon dioxide and metabolic water, as well as other gaseous wastes, in animals.

The kidney is the main organ involved in maintaining water balance and excreting harmful substances in mammals.

Elasmobranchs, marine birds, and some reptiles have a structure called a *salt gland* to secrete NaCl from their bodies. These animals require a lower internal NaCl concentration than the surrounding seawater, which causes a concentration gradient favouring the influx of salt. Therefore, they need a way to secrete it. The solution is provided by glands in the rectum of sharks and the skulls of marine birds and reptiles which produces a concentrated salt solution for secretion. The sodium ions are removed from the blood by these glands not by filtration, but by the sodium-transport mechanism (the sodium-potassium pump).

This Na^{+}/ K^{+}/ATPase activity allows for the movement of NaCl from the blood across the epithelium into the lumen of the salt gland for secretion. Interestingly, the shark rectal gland, bird nasal gland, fish gill, and the thick ascending Loop of Henle in the kidney all contain salt-secreting cells that transport NaCl by the same basic mechanism. Active transport produces an increase in the chloride concentration in the cytoplasm of epithelial cells. This results in the diffusion of chloride ions out of the cell across the apical surface. The build-up of chloride ions at the apical surface attracts sodium ions to diffuse between the cells (the paracellular route).

Insects have a network of Malpighian tubules extending throughout much of the body cavity and attached to the alimentary canal between the midgut and the hindgut. The secretory cells which line the walls of these long, thin tubules secrete KCl, NaCl, and phosphate from the hemolymph (blood) into the lumen of the tubule. Smaller molecules, such as water, amino acids, and sugars diffuse down their concentration gradient and into the lumen. The fluid then flows along the tubule and into the gut. As the fluid passes through the hindgut, water and valuable ions are transported back into the hemolymph, leaving behind a concentrated waste for excretion from the body.

Waste products generated in metabolic processes are often toxic, and therefore must be eliminated before they can harm the organism. The major metabolic wastes produced by animals include carbon dioxide, metabolic water, and nitrogenous wastes. Small aquatic organisms are able to get rid of wastes by simple diffusion across membranes. More complex animals with circulatory systems rely on kidneys to filter wastes out of the blood and eliminate them from the body.

Carbon dioxide and metabolic water produced in respiration easily diffuse into the environment from respiratory surfaces. Nitrogenous waste excretion is more difficult, yet necessary. Elevated ammonia levels in the body can lead to convulsions, coma, and even death. This is because ammonium ions can substitute for potassium ions in ion-exchange mechanisms. Ammonia can also adversely affect metabolism and amino acid transport. Excessive amounts of ammonia in the system elevates bodily pH, which causes changes in the tertiary structure of proteins, and thus cellular functions can be altered.

There are three main types of nitrogenous wastes: ammonia, urea, and uric acid. The type of waste an animal excretes depends on its living environment, because nitrogenous waste excretion is accompanied by a certain amount of water loss. Ammonotelic (ammonia-excreting) animals generally live only in aquatic habitats, because ammonia is extremely toxic, and a large volume of water is required to maintain the excreted ammonia level lower than the body level. This is needed because ammonia excretion relies on passive diffusion, so a gradient is required between the organism and the environment in order for the ammonia to flow from high concentration to low concentration.

Whereas most excretion of ammonia occurs across the gills of aquatic animals, mammals do excrete some ammonia in the urine. Amino groups are enzymatically transformed into glutamate, and then changed to glutamine in the liver. Glutamine can cross the kidney membranes (whereas amino acids can not). In the kidney tubules, the glutamine is deaminated to ammonia and then excreted in the urine.

Although ammonia excretion is present in some forms in mammals, the major nitrogenous waste excreted is urea. Urea is less toxic than ammonia, and requires less water for elimination. Therefore, ureotelic (urea-excreting) animals are most often (but not exclusively) terrestrial. A downside to urea excretion is that urea synthesis requires energy, in the form of ATP. Vertebrates synthesize urea in the liver using the ornithine-urea cycle. Teleosts and invertebrates produce urea from uric acid via the uricolytic pathway.

Birds, reptiles, and most terrestrial arthropods often are subject to very limited water availability, so even urea excretion is not possible. Therefore, these uricotelic (uric acid-excreting) animals synthesize uric acid, which requires even less water than urea for elimination. The ability to produce uric acid, which is relatively insoluble, is quite important to birds and reptiles prior to hatching. Nitrogenous wastes can be safely stored within the egg in the form of uric acid, whereas a build-up of either ammonia or urea would be deadly.

The kidney contains numerous functional units, called nephrons, which produce urine through a series of steps: glomerular filtration of the blood, tubular reabsorption of the glomerular filtrate, and tubular secretion of harmful substances.

GLOMERULAR FILTRATION

Blood flows from the afferent arteriole into the glomerulus, a tuft of fenestrated capillaries enclosed in the Bowman's capsule. Here 15 to 25 per cent of the plasma's water and solutes are filtered through a single-cell layer of the capillary walls, through a basement membrane, and into the lumen of the Bowman's capsule. The filtrate then flows into the renal tubule, to undergo tubular reabsorption.

The rate of glomerular filtration depends on three factors: the hydrostatic pressure difference between the capillaries and the Bowman's capsule (due to blood pressure), the colloid osmotic pressure, which opposes filtration, and the hydraulic permeability of the three-layered tissue separating the capillaries and the lumen of the Bowman's capsule.

Overall, blood pressure in the body has a major effect on the glomerular filtration rate, because the amount of blood passing through the glomerulus determines how much and how fast the fluid can be filtered.

Among the dangers of very low blood pressure, therefore, is the loss of kidney function. This is a primary reason for the use of inflatable "shock suits" on the lower body in cases of extreme blood loss from trauma. By reducing blood flow to the legs, blood pressure in the trunk is kept higher, in an effort to maintain kidney function.

The glomerular filtration rate can be regulated by the body through endocrine responses. In the case of autoregulation, increased blood pressure stretches walls of the afferent arteriole, which responds by contracting - thereby reducing fluctuation of blood pressure in the glomerulus.

A drop in blood pressure brings about a decrease in the glomerular filtration rate, which, in turn, results in a decrease in sodium ions in the filtrate. (The filtrate moves through the nephron more slowly, allowing more sodium to be reabsorbed along the way.). This lower sodium level in the filtrate is detected by the macula densa, modified cells of the wall of the distal convoluted tubule that lie adjacent to the afferent and efferent arterioles of the glomerulus.

In response to the low sodium, cells of the juxtaglomerular apparatus (JGA) release renin. This triggers a series of biochemical reactions which bring about an increase of blood pressure, and thereby an increase in GFR.

This series of reactions includes an increase in angiotensin II, which helps bring blood pressure back up by:

- Causing vasoconstriction in arterioles throughout much of the body, and
- Promoting increased synthesis of antidiiuretic hormone (adh), which increases resorption of water in the collecting ducts of the kidney, thereby increasing blood volume.

Angiotensin II also promotes the release of aldosterone from adrenal cortex, which promotes retention of both sodium and water, thereby helping to bring blood pressure back up.

The glomerular filtration rate can also be regulated by sympathetic nerve responses, which can result in the constriction or dilation of the afferent arterioles in times of great bodily stress. Vasoconstriction increases blood pressure, and therefore GFR, whereas vasodilation decreases blood pressure and GFR.

TUBULAR REABSORPTION

As the glomerular filtrate moves through the nephron, it changes composition dramatically. About 99% of the water in the original filtrate is reabsorbed, and less than 1% of the original NaCl content appears in the final urine.

How does this happen? First, the filtrate moves into the proximal tubule, where about 70% of the sodium ions are reabsorbed through active transport. Water and chloride ions follow passively. Glucose and amino acids are reabsorbed here. Approximately 75% of the glomerular filtrate is reabsorbed in the proximal tubule. This is aided by the so-called "brush border" of microvilli cells which function in increasing the surface area for reabsorption.

After moving through the proximal tubule, the filtrate moves on to the descending limb of the Loop of Henle. The cells in this area have no brush border, and there is no active salt transport here. The cells have a low permeability to urea and salt, but are very permeable to water. As the filtrate descends the Loop of Henle, water diffuses out because of the high salt concentration in the surrounding tissue. This is part of the urine-concentrating system of the nephron. The Loop of Henle acts as a countercurrent multiplier. For this reason, mammals living in marine or desert environments have longer loops of Henle and can conserve more water by producing a more concentrated urine.

The filtrate moves along the Loop of Henle to the thin segment of the ascending limb, which is highly permeable to Na^+ and Cl^-, and impermeable to water and urea. As the filtrate moves up the thin segment, Na^+ and Cl^- diffuse out because there is a higher concentration in the filtrate than in the surrounding tissues. The filtrate then moves to the medullary thick ascending limb, which is involved in the active transport of $Na^+$and Cl^- outward from the lumen into the interstitial space. This causes the fluid reaching the distal tubule to be hypoosmotic to the interstitial fluid, and allows for the passive transport of water out of the tubule.

The distal tubule functions in transporting K^+, H^+, and NH_3 into the lumen, and Na^+, Cl^-, and HCO_3^- out. Transport of salts in this area is under endocrine control and adjusted according to osmotic conditions.

The filtrate then moves into the collecting duct, which carries the fluid to the renal pelvis, to the ureters, and out of the body through the urethra. The epithelium of the collecting duct is permeable to water, but not salt or urea. The permeability o the collecting duct to water is controlled by the hormone ADH from the posterior pituitary gland. This response causes the filtrate, which is hypoosmotic to the interstitial fluid at this point, to lose water by osmosis and therefore increase the concentration of salts and urea in the urine. At the bottom of the collecting duct, the epithelium is permeable to urea. The diffusion of some urea out of the filtrate and into the surrounding tissue helps produce the interstitial concentration gradient necessary for the diffusion of water out of the descending limb of the Loop of Henle. The urea also helps draw water out of the filtrate passing down the collecting duct, thereby enabling the kidney to excrete urine that is hypertonic to the general body fluids, a property that is important in water conservation.

TUBULAR SECRETION

In several places along the nephron, substances that are not part of the initial filtrate (because the molecules are too big to be filtered through the glomerulus) are actively transported from the blood into the filtrate for elimination from the body in the urine. Some substances, such as toxins and drugs, are processed in the liver and conjugated with glucouronic acid. This marks them for removal from blood capillaries in the kidney and transport into the lumen of the nephron to become part of the filtrate.

Regulation of pH is governed by the carbon dioxide/ bicarbonate buffering system in the body, which consists of three steps:

$CO_2 + H_2O$ <==> H_2CO_3 <==> $HCO_3^- + H^+$

$CO_2 + OH^- + H^+$ <==> $HCO_3^- + H^+$

HOH <==> $OH^- + H^+$

The excretion of acid by the kidney is one of the two major factors which influence this system (the other being the excretion of carbon dioxide by the lungs). The excretion of hydrogen ions (acid) in the urine is primarily responsible for maintaining the plasma HCO_3^- concentration.

Mammalian urine is mildly acidic, with a pH of about 6, and contains no bicarbonate. However, the initial glomerular filtrate has a high bicarbonate concentration and a low hydrogen ion concentration. Therefore, in the process of urine formation, acid must be added to the filtrate, and bicarbonate must be removed. Therefore, the excretion of H^+ and the recovery of HCO_3^- are both important mechanisms by which the kidneys help the body regulate pH.

This process is accomplished by special cells in the distal tubule and collecting duct, called A-type cells and B-type cells. The A-type cells are acid-secreting cells that have a proton ATPase in the apical membrane and a Cl^-/ HCO_3^- exchange system in the basolateral membrane. The cells also contain carbonic anhydrase, which hydrates carbon dioxide passing through the membrane to form protons and bicarbonate ions. The protons formed are pumped back into the lumen and can react with the bicarbonate in the filtrate to form carbon dioxide and water, which can diffuse back into the cell, and create an uptake of bicarbonate back into the blood.

INTERACTION FORCES BETWEEN SOIL

On the basis of organic matter content, soils are characterized as mineral or organic. Mineral soils form most of the world's cultivated land and may contain from a trace to 30 per cent organic matter. Organic soils are naturally rich in organic matter principally for climatic reasons. Although they contain more than 30 per cent organic matter, it is precisely for this reason that they are not vital cropping soils.

This soils bulletin concentrates on the organic matter dynamics of cropping soils. In brief, it discusses circumstances that deplete organic matter and the negative outcomes of this. The bulletin then moves on to more proactive solutions. It reviews a "basket" of practices in order to show how they can increase organic matter content and discusses the land and cropping benefits that then accrue.

Soil organic matter is any material produced originally by living organisms (plant or animal) that is returned to the soil and goes through the decomposition process. At any given time, it consists of a range of materials from the intact original tissues of plants and animals to the substantially decomposed mixture of materials known as humus.

Fig. 3.1 Components of Soil Organic Matter and their Functions

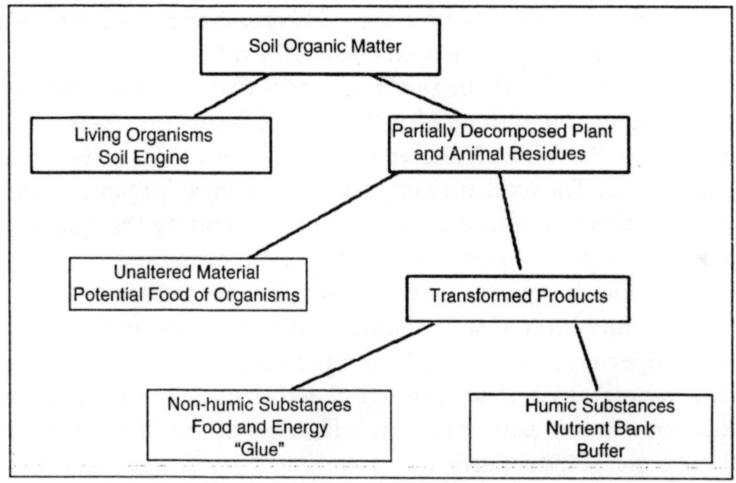

Fig. 3.2

Most soil organic matter originates from plant tissue. Plant residues contain 60-90 per cent moisture. The remaining dry matter consists of carbon (C), oxygen, hydrogen (H) and small amounts of sulphur (S), nitrogen (N), phosphorus (P), potassium (K), calcium (Ca) and magnesium (Mg). Although present in small amounts, these nutrients are very important from the viewpoint of soil fertility management.

Soil organic matter consists of a variety of components. These include, in varying proportions and many intermediate stages, an active organic fraction including microorganisms (10-40 per cent), and resistant or stable organic matter (40-60 per cent), also referred to as humus.

Forms and classification of soil organic matter have been described by Tate and Theng. For practical purposes, organic matter may be divided into aboveground and belowground fractions. Aboveground organic matter comprises plant residues and animal residues; belowground organic matter consists of living soil fauna and microflora, partially decomposed plant and animal residues, and humic substances. The C:N ratio is also used to indicate the type of material and ease of decomposition;

hard woody materials with a high C:N ratio being more resilient than soft leafy materials with a low C:N ratio.

Although soil organic matter can be partitioned conveniently into different fractions, these do not represent static end products. Instead, the amounts present reflect a dynamic equilibrium. The total amount and partitioning of organic matter in the soil is influenced by soil properties and by the quantity of annual inputs of plant and animal residues to the ecosystem. For example, in a given soil ecosystem, the rate of decomposition and accumulation of soil organic matter is determined by such soil properties as texture, pH, temperature, moisture, aeration, clay mineralogy and soil biological activities. A complication is that soil organic matter in turn influences or modifies many of these same soil properties.

Organic matter existing on the soil surface as raw plant residues helps protect the soil from the effect of rainfall, wind and sun. Removal, incorporation or burning of residues exposes the soil to negative climatic impacts, and removal or burning deprives the soil organisms of their primary energy source. Organic matter within the soil serves several functions.

From a practical agricultural standpoint, it is important for two main reasons:

1. As a "revolving nutrient fund"; and
2. As an agent to improve soil structure, maintain tilth and minimize erosion.

As a revolving nutrient fund, organic matter serves two main functions:

1. As soil organic matter is derived mainly from plant residues, it contains all of the essential plant nutrients. Therefore, accumulated organic matter is a storehouse of plant nutrients.
2. The stable organic fraction (humus) adsorbs and holds nutrients in a plant-available form.

Organic matter releases nutrients in a plant-available form upon decomposition. In order to maintain this nutrient cycling system, the rate of organic matter addition from crop residues, manure and any other sources must equal the rate of decomposition, and take into account the rate of uptake by plants and losses by leaching and erosion.

Where the rate of addition is less than the rate of decomposition, soil organic matter declines. Conversely, where the rate of addition is higher than the rate of decomposition, soil organic matter increases. The term steady state describes a condition where the rate of addition is equal to the rate of decomposition.

In terms of improving soil structure, the active and some of the resistant soil organic components, together with micro-organisms (especially fungi), are involved in binding soil particles into larger aggregates. Aggregation is important for good soil structure, aeration, water infiltration and resistance to erosion and crusting.

Traditionally, soil aggregation has been linked with either total C or organic C levels. More recently, techniques have developed to fractionate C on the basis of lability (ease of oxidation), recognizing that these subpools of C may have greater effect on soil physical stability and be more sensitive indicators than total C values of carbon dynamics in agricultural systems. The labile carbon fraction has been shown to be an indicator of key soil chemical and physical properties. For example, this fraction has been shown to be the primary factor controlling aggregate breakdown in Ferrosols (non-cracking red clays), measured by the percentage of aggregates measuring less than 0.125 mm in the surface crust after simulated rain in the laboratory.

The resistant or stable fraction of soil organic matter contributes mainly to nutrient holding capacity (cation exchange capacity [CEC]) and soil colour. This fraction of organic matter decomposes very slowly. Therefore, it has less influence on soil fertility than the active organic fraction.

4

Nonorganic Matter

SOIL ORGANIC MATTER

When plant residues are returned to the soil, various organic compounds undergo decomposition. Decomposition is a biological process that includes the physical breakdown and biochemical transformation of complex organic molecules of dead material into simpler organic and inorganic molecules.

The continual addition of decaying plant residues to the soil surface contributes to the biological activity and the carbon cycling process in the soil. Breakdown of soil organic matter and root growth and decay also contribute to these processes. Carbon cycling is the continuous transformation of organic and inorganic carbon compounds by plants and micro- and macro-organisms between the soil, plants and the atmosphere.

Decomposition of organic matter is largely a biological process that occurs naturally. Its speed is determined by three major factors: soil organisms, the physical environment and the quality of the organic matter. In the decomposition process, different products are released: carbon dioxide (CO_2), energy, water, plant nutrients and resynthesized organic carbon compounds. Successive decomposition of dead material and modified organic matter results in the formation of a more complex organic matter called humus. This process is called humification. Humus affects soil properties. As it slowly decomposes, it colours the soil darker; increases soil aggregation and aggregate stability; increases the CEC (the ability to attract and retain nutrients); and contributes N, P and other nutrients.

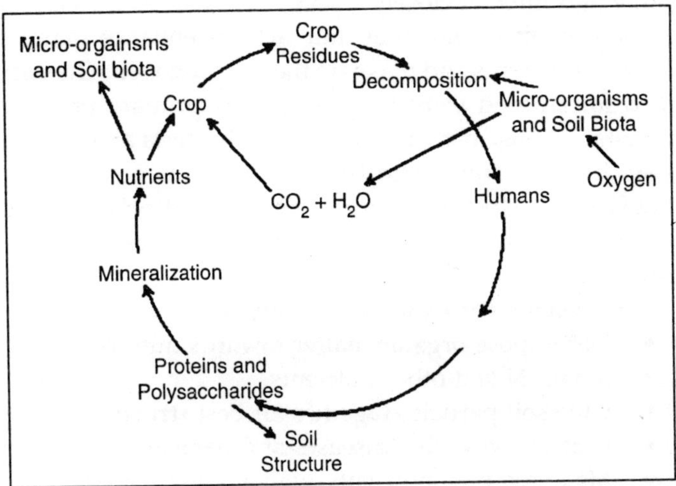

Fig. 4.1 Carbon Cycle

Soil organisms, including micro-organisms, use soil organic matter as food. As they break down the organic matter, any excess nutrients (N, P and S) are released into the soil in forms that plants can use. This release process is called mineralization. The waste products produced by micro-organisms are also soil organic matter. This waste material is less decomposable than the original plant and animal material, but it can be used by a large number of organisms. By breaking down carbon structures and rebuilding new ones or storing the C into their own biomass, soil biota plays the most important role in nutrient cycling processes and, thus, in the ability of a soil to provide the crop with sufficient nutrients to harvest a healthy product. The organic matter content, especially the more stable humus, increases the capacity to store water and store (sequester) C from the atmosphere.

THE SOIL FOOD

The soil ecosystem can be defined as an interdependent life-support system composed of air, water, minerals, organic matter, and macro- and micro-organisms, all of which function together and interact closely.

The organisms and their interactions enhance many soil ecosystem functions and make up the soil food web. The energy needed for all food webs is generated by primary producers: the plants, lichens, moss, photosynthetic bacteria and algae that use sunlight to transform CO_2 from the atmosphere into carbohydrates. Most other organisms depend on the primary producers for their energy and nutrients; they are called consumers.

Some functions of a healthy soil ecosystem:

- Decompose organic matter towards humus.
- Retain N and other nutrients.
- Glue soil particles together for best structure.
- Protect roots from diseases and parasites.
- Make retained nutrients available to the plant.
- Produce hormones that help plants grow.
- Retain water.

Soil life plays a major role in many natural processes that determine nutrient and water availability for agricultural productivity.

The primary activities of all living organisms are growing and reproducing. By-products from growing roots and plant residues feed soil organisms. In turn, soil organisms support plant health as they decompose organic matter, cycle nutrients, enhance soil structure and control the populations of soil organisms, both beneficial and harmful (pests and pathogens) in terms of crop productivity.

The living part of soil organic matter includes a wide variety of micro-organisms such as bacteria, viruses, fungi, protozoa and algae. It also includes plant roots, insects, earthworms, and larger animals such as moles, mice and rabbits that spend part of their life in the soil.

The living portion represents about 5 per cent of the total soil organic matter. Micro-organisms, earthworms and insects help break down crop residues and manures by ingesting them and mixing them with the minerals in the soil, and in the process recycling energy and plant nutrients. Sticky substances on the skin of earthworms and those produced by fungi and bacteria help bind particles together.

Earthworm casts are also more strongly aggregated (bound together) than the surrounding soil as a result of the mixing of organic matter and soil mineral material, as well as the intestinal mucus of the worm. Thus, the living part of the soil is responsible for keeping air and water available, providing plant nutrients, breaking down pollutants and maintaining the soil structure.

The composition of soil organisms depends on the food source (which in turn is season dependent). Therefore, the organisms are neither uniformly distributed through the soil nor uniformly present all year.

However, in some cases their biogenic structures remain. Each species and group exists where it can find appropriate food supply, space, nutrients and moisture. Organisms occur wherever organic matter occurs.

Therefore, soil organisms are concentrated: around roots, in litter, on humus, on the surface of soil aggregates and in spaces between aggregates. For this reason, they are most prevalent in forested areas and cropping systems that leave a lot of biomass on the surface.

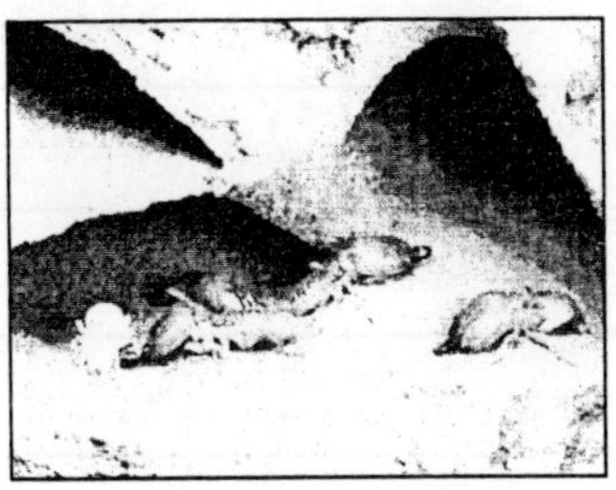

Fig. 4.2

The activity of soil organisms follows seasonal as well as daily patterns. Not all organisms are active at the same time. Most are barely active or even dormant. Availability of food is an important factor that influences the level of activity of soil organisms and thus is related to land use and management (Figure 4.3). Practices that increase numbers and activity of soil organisms include: no tillage or minimal tillage; and the maintenance of plant and annual residues that reduce disturbance of soil organisms and their habitat and provide a food supply.

DECOMPOSITION PROCESS

Fresh residues consist of recently deceased micro-organisms, insects and earthworms, old plant roots, crop residues, and recently added manures. Crop residues contain mainly complex carbon compounds originating from cell walls (cellulose, hemicellulose, etc.).

Chains of carbon, with each carbon atom linked to other carbons, form the "backbone" of organic molecules. These carbon chains, with varying amounts of attached oxygen, H, N, P and S, are the basis for both simple sugars and amino acids and more complicated molecules of long carbon chains or rings. Depending on their chemical structure, decomposition is rapid (sugars, starches and proteins), slow (cellulose, fats, waxes and resins) or very slow (lignin).

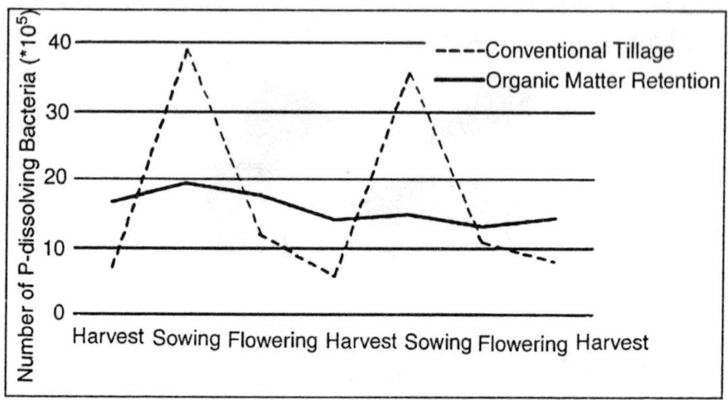

Fig. 4.3 Fluctuations in Microbial Biomass at Different Stages of Crop Development in Conventional Agriculture Compared with Systems with Residue Retention and High Organic Matter Input

Table. Essential Functions Performed by Different Members of Soil Organisms (Biota)

Functions	Organisms involved
Maintenance of soil structure	Bioturbating invertebrates and plant roots, mycorrhizae and some other micro-organisms
Regulation of soil hydrological processes	Most bioturbating invertebrates and plant roots
Gas exchange and carbon sequestration (accumulation in soil)	Mostly micro-organisms and plant roots, some C protected in large compact biogenic invertebrate aggregates
Soil detoxification	Mostly micro-organisms
Nutrient cycling	Mostly micro-organisms and plant roots, some soil- and litter-feeding invertebrates
Decomposition of organic matter	Various saprophytic and litter-feeding invertebrates (detritivores), fungi, bacteria, actinomycetes and other micro-organisms
Suppression of pests, parasites and diseases	Plants, mycorrhizae and other fungi, nematodes, bacteria and various other micro-organisms, collembola, earthworms, various predators
Sources of food and medicines	Plant roots, various insects (crickets, beetle larvae, ants, termites), earthworms, vertebrates, micro-organisms and their by-products
Symbiotic and asymbiotic relationships with plants and their roots	Rhizobia, mycorrhizae, actinomycetes, diazotrophic bacteria and various other rhizosphere micro-organisms, ants
Plant growth control (positive and negative)	Direct effects: plant roots, rhizobia, mycorrhizae, actinomycetes, pathogens, phytoparasitic nematodes, rhizophagous insects, plant-growth promoting rhizosphere micro-organisms, biocontrol agents Indirect effects: most soil biota

Table. Classification of Soil Organisms

Micro-organisms	Microflora	<5 μm	Bacteria Fungi
	Microfauna	<100 μm	Protozoa Nematodes
Macro-organisms	Meso-organisms	100 μm - 2 mm	Springtails Mites
	Macro-organisms	2 - 20 mm	Earthworms Millipedes Woodlice Snails and slugs
Plants	Algae	10 μm	
	Roots	> 10 μm	

During the decomposition process, microorganisms convert the carbon structures of fresh residues into transformed carbon products in the soil. There are many different types of organic molecules in soil. Some are simple molecules that have been synthesized directly from plants or other living organisms. These relatively simple chemicals, such as sugars, amino acids, and cellulose are readily consumed by many organisms. For this reason, they do not remain in the soil for a long time. Other chemicals such as resins and waxes also come directly from plants, but are more difficult for soil organisms to break down.

Humus is the result of successive steps in the decomposition of organic matter. Because of the complex structure of humic substances, humus cannot be used by many micro-organisms as an energy source and remains in the soil for a relatively long time.

NON-HUMIC SUBSTANCES: SIGNIFICANCE AND FUNCTION

Non-humic organic molecules are released directly from cells of fresh residues, such as proteins, amino acids, sugars, and starches. This part of soil organic matter is the active, or easily decomposed, fraction. This active fraction is influenced strongly by weather conditions, moisture status of the soil, growth stage of the vegetation, addition of organic residues, and cultural practices, such as tillage. It is the main food supply for various organisms in the soil.

Carbohydrates occur in the soil in three main forms: free sugars in the soil solution, cellulose and hemicellulose; complex polysaccharides; and polymeric molecules of various sizes and shapes that are attached strongly to clay colloids and humic substances. The simple sugars, cellulose and hemicellulose, may constitute 5-25 per cent of the organic matter in most soils, but are easily broken down by micro-organisms.

Polysaccharides (repeating units of sugar-type molecules connected in longer chains) promote better soil structure through their ability to bind inorganic soil particles into stable aggregates. Research indicates that the heavier polysaccharide molecules may be more important in promoting aggregate stability and water infiltration than the lighter molecules. Some sugars may stimulate seed germination and root elongation. Other soil properties affected by polysaccharides include CEC, anion retention and biological activity. The soil lipids form a very diverse group of materials, of which fats, waxes and resins make up 2-6 per cent of soil organic matter. The significance of lipids arises from the ability of some compounds to act as growth hormones. Others may have a depressing effect on plant growth.

Soil N occurs mainly (> 90 per cent) in organic forms as amino acids, nucleic acids and amino sugars. Small amounts exist in the form of amines, vitamins, pesticides and their degradation products, etc. The rest is present as ammonium (NH_4^-) and is held by the clay minerals.

COMPOUNDS AND FUNCTION OF HUMUS

Humus or humified organic matter is the remaining part of organic matter that has been used and transformed by many different soil organisms. It is a relatively stable component formed by humic substances, including humic acids, fulvic acids, hymatomelanic acids and humins. It is probably the most widely distributed organic carbon-containing material in terrestrial and aquatic environments. Humus cannot be decomposed readily because of its intimate interactions with soil mineral phases and is chemically too complex to be used by most organisms.

One of the most striking characteristics of humic substances is their ability to interact with metal ions, oxides, hydroxides, mineral and organic compounds, including toxic pollutants, to form water-soluble and water-insoluble complexes. Through the formation of these complexes, humic substances can dissolve, mobilize and transport metals and organics in soils and waters, or accumulate in certain soil horizons.

This influences nutrient availability, especially those nutrients present at microconcentrations only. Accumulation of such complexes can contribute to a reduction of toxicity, *e.g.* of aluminium (Al) in acid soils, or the capture of pollutants - herbicides such as Atrazine or pesticides such as Tefluthrin - in the cavities of the humic substances.

Humic and fulvic substances enhance plant growth directly through physiological and nutritional effects. Some of these substances function as natural plant hormones and are capable of improving seed germination, root initiation, uptake of plant nutrients and can serve as sources of N, P and S. Indirectly, they may affect plant growth through modifications of physical, chemical and biological properties of the soil, for example, enhanced soil water holding capacity and CEC, and improved tilth and aeration through good soil structure.

About 35-55 per cent of the non-living part of organic matter is humus. It is an important buffer, reducing fluctuations in soil acidity and nutrient availability. Compared with simple organic molecules, humic substances are very complex and large, with high molecular weights.

The characteristics of the well-decomposed part of the organic matter, the humus, are very different from those of simple organic molecules. While much is known about their general chemical composition, the relative significance of the various types of humic materials to plant growth is yet to be established.

HUMUS CONSISTS OF DIFFERENT HUMIC SUBSTANCES

- *Fulvic acids*: The fraction of humus that is soluble in water under all pH conditions. Their colour is commonly light yellow to yellow-brown.
- *Humic acids*: The fraction of humus that is soluble in water, except for conditions more acid than pH 2. Common colours are dark brown to black.
- *Humin*: The fraction of humus that is not soluble in water at any pH and that cannot be extracted with a strong base, such as sodium hydroxide (NaOH). Commonly black in colour.

The term acid is used to describe humic materials because humus behaves like weak acids. Fulvic and humic acids are complex mixtures of large molecules. Humic acids are larger than fulvic acids. Research suggests that the different substances are differentiated from each other on the basis of their water solubility. Fulvic acids are produced in the earlier stages of humus formation. The relative amounts of humic and fulvic acids in soils vary with soil type and management practices. The humus of forest soils is characterized by a high content of fulvic acids, while the humus of agricultural and grassland areas contains more humic acids.

NATURAL FACTORS

The transformation and movement of materials within soil organic matter pools is a dynamic process influenced by climate, soil type, vegetation and soil organisms. All these factors operate within a hierarchical spatial scale. Soil organisms are responsible for the decay and cycling of both macronutrients and micronutrients, and their activity affects the structure, tilth and productivity of the soil. In natural humid and subhumid forest ecosystems without human disturbance, the living and non-living components are in dynamic equilibrium with each other. The litter on the soil surface beneath different canopy layers and high biomass production generally result in high biological activity in the soil and on the soil surface.

Mollison and Slay distinguished the following five mechanisms:

1. A continuous soil cover of living plants, which together with the soil architecture facilitates the capture and infiltration of rainwater and protects the soil;
2. A litter layer of decomposing leaves or residues providing a continuous energy source for macro- and micro-organisms;
3. The roots of different plants distributed throughout the soil at different depths permit an effective uptake of nutrients and an active interaction with microorganisms;
4. The major period of nutrient release by micro-organisms coincides with the major period of nutrient demand by plants;
5. Nutrients recycled by deep-rooting plants and soil macrofauna and microfauna.

This equilibrium creates almost closed-cycle transfers of nutrients between soil and the vegetation adapted to such site conditions, resulting in almost perfect physical and hydric conditions for plant growth, *i.e.* a cool microclimate, increased evapotranspiration, good rooting conditions with good porosity and sufficient soil moisture. This facilitates water infiltration and prevents erosion and run-off. Thus, it results in clean water in the streams emanating from the area, a relatively smooth variation in streamflow during the year, and recharge of groundwater.

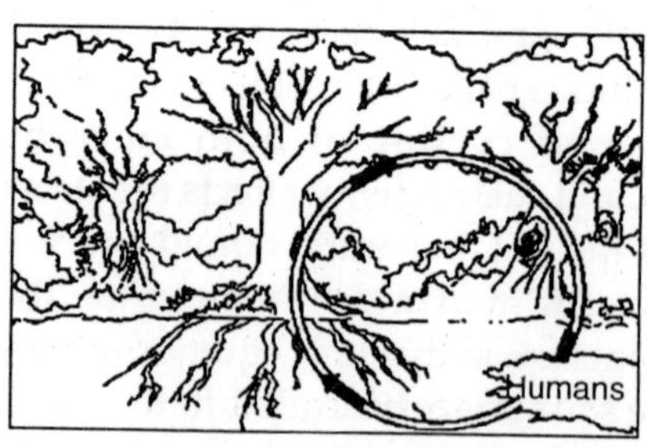

Fig. 4.4 Closed Cycle

In human-managed systems, the soil biological activity is influenced by the land use system, plant types and the management practices. The environmental and edaphic factors that control the activity of soil biota, and thus the balance between accumulation and decomposition of organic matter in the soil, are described below.

TEMPERATURE

Several field studies have shown that temperature is a key factor controlling the rate of decomposition of plant residues. Decomposition normally occurs more rapidly in the tropics than in temperate areas. Ladd and Amato reported that, despite differences in plant material and climate patterns, the decomposition of leguminous materials in southern Australian sites followed the same pattern as that of ryegrass for sites in Nigeria and the United Kingdom, although the time scales were different. Reaction rates doubled for each increase of 8-9 °C in the mean annual air temperature.

The relatively faster rate of decomposition induced by the continuous warmth in the tropics implies that high equilibrium levels of organic matter are difficult to achieve in tropical agro-ecosystems. Hence, large annual rates of organic inputs are needed to maintain an adequate labile soil organic matter pool in cultivated soils. Soils in cooler climates commonly have more organic matter because of slower mineralization (decomposition) rates.

SOIL MOISTURE AND WATER SATURATION

Soil organic matter levels commonly increase as mean annual precipitation increases. Conditions of elevated levels of soil moisture result in greater biomass production, which provides more residues, and thus more potential food for soil biota.

Soil biological activity requires air and moisture. Optimal microbial activity occurs at near "field capacity", which is equivalent to 60-per cent water-filled pore space. On the other hand, periods of water saturation lead to poor aeration. Most soil organisms need oxygen, and thus a reduction of oxygen in the soil leads to a reduction of the mineralization rate as these organisms become inactive or even die. Some of the transformation processes become anaerobic, which can lead to damage to plant roots caused by waste products or favourable conditions for disease-causing organisms. Continued production and slow decomposition can lead to very large organic matter contents in soils with long periods of water saturation (*e.g.* peat soils, and tea crops in India).

With the exception of the hyperhumid regions, the climates of vast areas of the humid, subhumid and semi-arid tropics are characterized by distinct wet and dry seasons. In the wet-dry tropics, large amounts of nitrate often occur in the surface soil during the first part of the rainy season. This accelerated nitrogen mineralization caused by a large increase in microbial activity is the result of the first few rains activating the labile soil organic matter.

Farmers who practise "slash and burn" agriculture often choose early planting in order to take advantage of this flush of inorganic N before it is lost through leaching and run-off. In these low-input systems, the amount of nitrate present in the soil during the early part of the rainy season is related closely to the organic matter content of the soil. N availability diminishes during the later part of the rainy season.

SOIL TEXTURE

Soil organic matter tends to increase as the clay content increases. This increase depends on two mechanisms. First, bonds between the surface of clay particles and organic matter retard the decomposition process. Second, soils with higher clay content increase the potential for aggregate formation. Macroaggregates physically protect organic matter molecules from further mineralization caused by microbial attack. For example, when earthworm casts and the large soil particles they contain are split by the joint action of several factors (climate, plant growth and other organisms), nutrients are released and made available to other components of soil micro-organisms.

Under similar climate conditions, the organic matter content in fine textured (clayey) soils is two to four times that of coarse textured (sandy) soils.

Kaolinite, the main clay mineral in many upland soils in the tropics, has a much smaller specific surface and nutrient exchange capacity than most other clay minerals. Therefore, kaolinitic soils contain considerably fewer clay-humus complexes. In addition, the unprotected labile humic substances are vulnerable to decomposition under appropriate soil moisture conditions. Thus, high levels of organic matter are difficult to maintain in cultivated kaolinitic soils in the wet-dry tropics, because climate and soil conditions favour rapid decomposition. In contrast, organic matter can persist as organo-oxide complexes in soils rich in iron and aluminium oxides. Such properties favour the formation of soil microaggregates, typical of many fine-textured, oxide-rich, high base-status soils in the tropics. These soils are known for their low bulk density, high microporosity, and high organic-matter retention under natural vegetation, but also for their high phosphate fixation capacity on the oxides when used for crop production. Current knowledge suggests that whereas organic matter contributes to the dark colour of Vertisols, it is not considered important in determining either the development, robustness or resilience of structure in these soils. Organic matter levels tend to be low in Vertisols; even as low as 10 g/ kg.

Parent material influences organic matter accumulation not only through its effect on soil texture. Soils developed from inherently rich material, such as basalt, are more fertile than soils formed from granitic material, which contains less mineral nutrients. Moreover, the former experience more organic matter accumulation because of abundant vegetative growth.

TOPOGRAPHY

Organic matter accumulation is often favoured at the bottom of hills. There are two reasons for this accumulation: conditions are wetter than at mid- or upper-slope positions, and organic matter is transported to the lowest point in the landscape through run-off and erosion. Similarly, soil organic matter levels are higher on northfacing slopes (in the Northern Hemisphere) compared with south-facing slopes (and the other way around in the Southern Hemisphere) because temperatures are lower.

SALINITY AND ACIDITY

Salinity, toxicity and extremes in soil pH (acid or alkaline) result in poor biomass production and, thus in reduced additions of organic matter to the soil. For example, pH affects humus formation in two ways: decomposition, and biomass production. In strongly acid or highly alkaline soils, the growing conditions for micro-organisms are poor, resulting in low levels of biological oxidation of organic matter. Soil acidity also influences the availability of plant nutrients and thus regulates indirectly biomass production and the available food for soil biota. Fungi are less sensitive than bacteria to acid soil conditions.

VEGETATION AND BIOMASS PRODUCTION

The rate of soil organic matter accumulation depends largely on the quantity and quality of organic matter input. Under tropical conditions, applications of readily degradable materials with low C:N ratios, such as green manure and leguminous cover crops, favour decomposition and a short-term increase in the labile nitrogen pool during the growing season. On the other hand, applications of plant materials with both large C:N ratios and lignin contents such as cereal straw and grasses generally favour nutrient immobilization, organic matter accumulation and humus formation, with increased potential for improved soil structure development.

Plant constituents such as lignin and other polyphenols retard decomposition. In an experiment in southern Nigeria to compare management effects on soil organic matter accumulation, a three-year fallow with Guinea grass (*Panicum maximum*), which has a high lignin content, maintained a carbon level comparable to that under forest fallow. However, fallowing with leguminous species such as pigeon pea (*Cajanus *cajan*) caused a significant decline in soil total C. Palm and Sanchez reported that both the decomposition rate and the N-release patterns of three tropical legumes were related to the amount of polyphenol compounds such as lignin in the leaf. *Erythrina* leaves had the lowest concentrations of polyphenols and the fastest decomposition rate of the three species studied.

Root turnover also constitutes an important addition of humus into the soil, and consequently it is important for carbon sequestration. In forests, most organic matter is added as superficial litter. However, in grassland ecosystems, up to two-thirds of organic matter is added through the decay of roots.

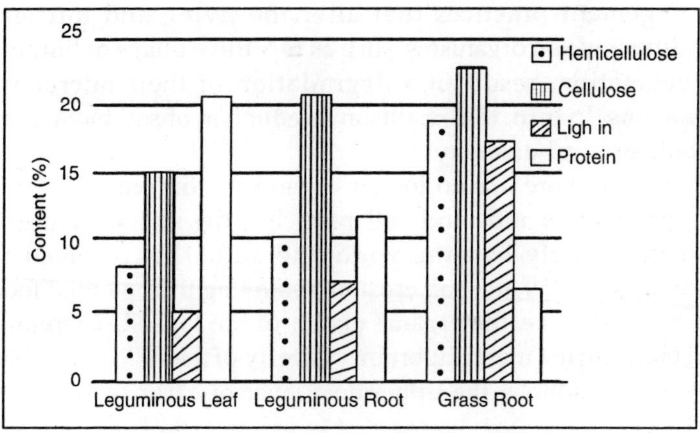

Fig. 4.5 Composition of Leaves and Roots of Leguminous and Grass Species

PRACTICES THAT INFLUENCE THE AMOUNT OF ORGANIC MATTER

HUMAN INTERVENTIONS

Various types of human activity decrease soil organic matter contents and biological activity. However, increasing the organic matter content of soils or even maintaining good levels requires a sustained effort that includes returning organic materials to soils and rotations with high-residue crops and deep- or dense-rooting crops. It is especially difficult to raise the organic matter content of soils that are well aerated, such as coarse sands, and soils in warm-hot and arid regions because the added materials decompose rapidly. Soil organic matter levels can be maintained with less organic residue in finetextured soils in cold temperate and moist-wet regions with restricted aeration.

Practices that Decrease Soil Organic Matter

Any form of human intervention influences the activity of soil organisms and thus the equilibrium of the system. Management practices that alter the living and nutrient conditions of soil organisms, such as repetitive tillage or burning of vegetation, result in a degradation of their microenvironments. In turn, this results in a reduction of soil biota, both in biomass and diversity.

Where there are no longer organisms to decompose soil organic matter and bind soil particles, the soil structure is damaged easily by rain, wind and sun. This can lead to rainwater run-off and soil erosion, removing the potential food for organisms, *i.e.* the organic matter of the topsoil. Therefore, soil biota are the most important property of the soil, and "when devoid of its biota, the uppermost layer of earth ceases to be soil".

Fig. 4.6 Open Cycle System

The factors leading to reduction in soil organic matter in an open cycle system can be grouped as factors that result in:

- A decrease in biomass production;

- A decrease in organic matter supply;
- Increased decomposition rates.

Fig. 4.7

DECREASE IN BIOMASS PRODUCTION

REPLACEMENT OF PERENNIAL VEGETATION

A consequence of clearing forest for agriculture is the disappearance of the litter layer, with a consequent reduction in the numbers and variety of soil organisms.

While many temperate forest species appear to adapt well to grassland, the effects of deforestation in the tropics appear to be more marked.

Studies have shown that as soil biodiversity declines, adapted species may take over from the indigenous species and the composition may change drastically.

Soil macrofaunal biomass and population density fell to 6 and 17 per cent, respectively, in cultivated plots, compared with primary forest in Peruvian Amazonia. In Suriname, the number of animals per square metre has fallen to 36 per cent and the diversity of species has fallen to 28 per cent compared with primary forest.

The indigenous species have largely disappeared, but adapted species have been available for recolonization. The composition of the macrofaunal community has changed drastically.

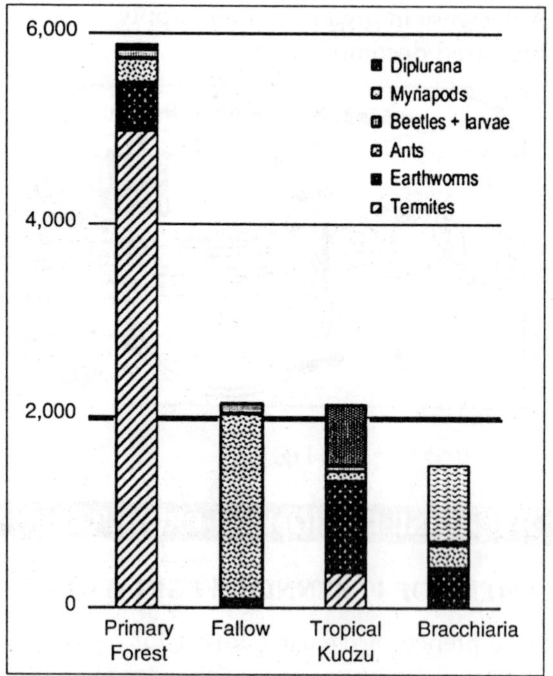

Fig. 4.8 Composition of Soil Macrofauna Under Primary Forest, Fallow, Kudzu and Grass Vegetation

Fig. 4.9

REPLACEMENT OF MIXED VEGETATION WITH MONOCULTURE OF CROPS AND PASTURES

The simplification of vegetation and the disappearance of the litter layer under grassland and monocrop production systems lead to a decrease in faunal diversity. Although root systems (especially of grasses) can be extensive and explore vast areas of soil, the root exudates from one single crop will attract only a few different microbial species. This in turn will affect the predator diversity.

The more opportunistic pathogen species will be able to acquire space near the crop and cause harm. Continuous cultivation and grazing also leads to compaction of soil layers, which in turn affects the circulation of air. Anaerobic conditions in the soil stimulate the growth of different micro-organisms, resulting in more pathogenic organisms.

HIGH HARVEST INDEX

One of the consequences of the green revolution was the replacement of indigenous varieties of species with high-yielding varieties (HYVs). These HYVs often produce more grain and less straw, compared with locally developed varieties; the harvest index of the crop (ratio of grain to total plant mass aboveground) is increased. From a production point of view, this is a logical approach. However, this is less desirable from a conservation point of view. Reduced amounts of crop residues remain after harvest for soil cover and organic matter, or for grazing of livestock (which results in manure). Moreover, where animals graze the residues, even less remains for conservation purposes.

USE OF BARE FALLOW

Traditionally, a fallow period is used after a period of crop production to give the land some "rest" and to regenerate its original state of productivity. Usually, this is necessary in production systems that have drawn down the nutrient supply and altered the soil biota significantly, such as in slash-and-burn systems or conventional tillage systems.

Some farmers use bare fallow to regenerate their lands. However, apart from spontaneous weed growth, this means there is no energy source for the soil biota present on the land. Instead of recovering the soil food web, the soil organic matter is degraded further and the lack of cover can result in severe erosion and run-off when the rains start after the dry season.

Fig. 4.10

DECREASE IN ORGANIC MATTER SUPPLY

BURNING OF NATURAL VEGETATION AND CROP RESIDUES

The burning of maize, rice and other crop residues in the field is a common practice. Residues are usually burned to help control insects or diseases or to make fieldwork easier in the following season. Burning destroys the litter layer and so diminishes the amount of organic matter returned to the soil. The organisms that inhabit the surface soil and litter layer are also eliminated. For future decomposition to take place, energy has to be invested first in rebuilding the microbial community before plant nutrients can be released. Similarly, fallow lands and bush are burned before cultivation. This provides a rapid supply of P to stimulate seed germination. However, the associated loss of nutrients, organic matter and soil biological activity has severe long-term consequences.

OVERGRAZING

There is a tendency throughout the world to overstock grazing land above its carrying capacity. Cows, draught animals and small ruminants graze on communal grazing areas and on roadsides, stream banks and other public land. Overgrazing destroys the most palatable and useful species in the plant mixture and reduces the density of the plant cover, thereby increasing the erosion hazard and reducing the nutritive value and the carrying capacity of the land.

REMOVAL OF CROP RESIDUES

Many farmers remove residues from the field for use as animal feed and bedding or to make compost. Later, these residues return to contribute to soil fertility as manures or composts. However, residues are sometimes removed from the field and not returned. This removal of plant material impoverishes the soil as it is no longer possible to recycle the plant nutrients present in the residues.

Fig. 4.11

INCREASED DECOMPOSITION RATES

TILLAGE PRACTICES

Tillage is one of the major practices that reduces the organic matter level in the soil. Each time the soil is tilled, it is aerated. As the decomposition of organic matter and the liberation of C are aerobic processes, the oxygen stimulates or speeds up the action of soil microbes, which feed on organic matter.

This means that:

- When ploughed, the residues are incorporated in the soil together with air and come into contact with many micro-organisms, which accelerates the carbon cycle. The decomposition is faster, resulting in the formation of less stable humus and an increased liberation of CO_2 to the atmosphere, and thus a reduction in organic matter.

- The residues on the soil surface slow the carbon cycle because they are exposed to fewer micro-organisms and thus wane more slowly, resulting in the production of humus (which is more stable), and liberating less CO_2 to the atmosphere.

Fig. 4.12

Table. Tillage Induced Flush of Decomposition of Organic Matter

Type of tillage	Organic matter lost in 19 days (kg/ha)
Mouldboard plough + disc harrow (2x)	4 300
Mouldboard plough	2 230
Disc harrow	1 840
Chisel plough	1 720
Direct seeding	860

In terms of short-term organic matter loss, the more a soil is tilled, the more the organic matter is broken down. There are also longer-term losses, attributed to repeated, annual cultivation. Cropping systems that return little residue to the soil accelerate this decline. Many modern cropping systems combine frequent tillage with small amounts of residue, with resultant reductions in the organic matter content of many soils. Historically, manure application (from farm livestock) was common, and it was a dynamic way of maintaining organic matter levels despite repeated cultivation and low residue returns to the soil. Increased on-farm mechanization has reduced livestock numbers, so this source of organic material has been reduced considerably.

Organic matter production and conservation is affected dramatically by conventional tillage, which not only decreases soil organic matter but also increases the potential for erosion by wind and water.

The impact occurs in many ways:

- Ploughing leaves no residues on the soil surface to lessen the impact of rain.
- Ploughing reduces the quantity of food sources for earthworms and disturbs their burrows and living space, hence populations of certain species decrease drastically. Moreover, reduction of earthworm numbers reduces their impact, through burrowing, in increasing porosity and aeration (particularly continuous macropores) and lowers their ability to bury and incorporate plant residues, which facilitates rapid decomposition of organic matter.
- Tillage by repeated hoeing or discing smoothes the surface and destroys natural soil aggregates and channels that connect the surface with the subsoil, leaving the soil susceptible to erosion. Old root channels and earthworm holes are eliminated, as are the cracks between natural aggregates. The large pores, the ones destroyed by conventional tillage practices, are necessary to conduct water into the soil during rainfall.
- The development of a plough pan or hoe pan, a layer of compacted soil resulting from smearing action at the bottom of the plough or hoe, may retard both root penetration and water infiltration.
- Ploughing or discing under dry conditions exacerbates the pulverization of the soil, causing the soil surface to crust more easily, leading to greater water run-off

and erosion. This is exacerbated by reduced soil surface roughness, which leaves few depressions for temporary storage of water during intense storms.

- Increased run-off during rainstorms may also increase the possibility of drought stress later in the season, because water that runs off the field does not infiltrate into the soil to remain available to plants.

In some circumstances, imbalances of certain soil organisms can disrupt soil structure and processes, *e.g.* certain earthworm species in rice fields or pastures.

Fig. 4.13

DRAINAGE

Decomposition of organic matter occurs more slowly in poorly aerated soils, where oxygen is limiting or absent, compared with well-aerated soils. For this reason, organic matter accumulates in wet soil environments. Soil drainage is determined strongly by topography - soils in depressions at the bottom of hills tend to remain wet for extended periods of time because they receive water (and sediments) from upslope.

Soils may also have a layer in the subsoil that inhibits drainage, again exacerbating waterlogging and reduction in organic matter decomposition. In a permanently waterlogged soil, one of the major structural parts of plants, lignin, does not decompose at all. The ultimate consequence of extremely wet or swampy conditions is the development of organic (peat or muck) soils, with organic matter contents of more than 30 per cent. Where soils are drained artificially for agricultural or other uses, the soil organic matter decomposes rapidly.

FERTILIZER AND PESTICIDE USE

Initially, the use of fertilizer and pesticides enhances crop development and thus production of biomass (especially important on depleted soils). However, the use of some fertilizers, especially N fertilizers, and pesticides can boost micro-organism activity and thus decomposition of organic matter. The chemicals provide the microorganisms with easy-to-use N components. This is especially important where the C: N ratio of the soil organic matter is high and thus decomposition is slowed by a lack of N.

PRACTICES THAT INCREASE SOIL ORGANIC MATTER

Increased concern about the environmental and economic impacts of conventional crop production has stimulated interest in alternative systems. Central to such systems is the need to promote and maintain soil biological processes and minimize fossil fuel inputs in the form of fertilizers, pesticides and mechanical cultivation. All activities aimed at the increase of organic matter in the soil help in creating a new equilibrium in the agro-ecosystem.

For a system of natural resource management to be balanced, and thus sustainable, it must be able to withstand sharp climatic fluctuations, and to evolve steadily in response to social changes and changes in the costs and availability of inputs of land, labour and knowledge. The more diverse and complex an agricultural system is, the more stable and sustainable it will be in the face of unpredictable vagaries of climate and market. Thus, annual crops, woody perennials and nonwoody perennials may be combined in various ways with livestock or trees, or both, in what are now commonly called agrosilvipastoral systems. Different approaches are required for different soil and climate conditions. However, the activities will be based on the same principle: increasing biomass production in order to build active organic matter. Active organic matter provides habitat and food for beneficial soil organisms that help build soil structure and porosity, provide nutrients to plants, and improve the water holding capacity of the soil. Several cases have demonstrated that it is possible to restore organic matter levels in the soil (Figure 4.14). Activities that promote the accumulation and supply of organic matter, such as the use of cover crops and refraining from burning, and those that reduce decomposition rates, such as reduced and zero tillage, lead to an increase in the organic matter content in the soil.

Ways to increase organic matter contents of soils:

- Compost
- Cover crops/green manure crops
- Crop rotation
- Perennial forage crops
- Zero or reduced tillage
- Agroforestry

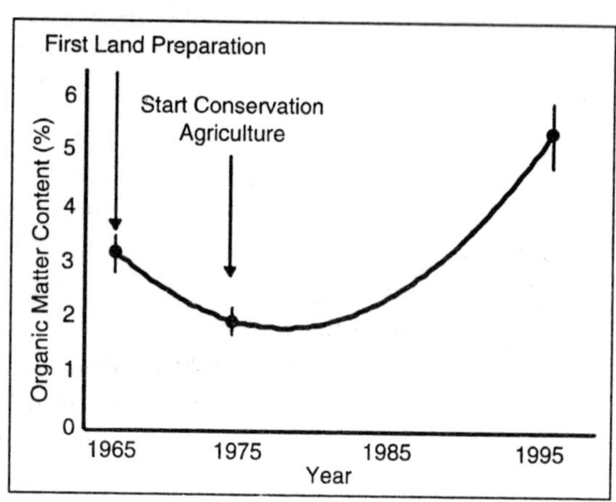

Fig. 4.14 Evaluation of the Organic Matter Content of a Soil in Paraná

INCREASED BIOMASS PRODUCTION

INCREASED WATER AVAILABILITY FOR PLANTS: WATER HARVESTING AND IRRIGATION

In dry conditions, water may be provided through

irrigation or water harvesting. The increased water availability enhances biomass production, soil biological activity and plant residues and roots that provide organic matter.

The concept of water harvesting includes various technologies for run-off management and utilization. It involves capture of run-off (in some cases through treating the upstream capture area), and its concentration on a runon area for use by a specific crop (annual or perennial) in order to enhance crop growth and yields, or its collection and storage for supplementary irrigation or domestic or livestock purposes. The objective of designing a water harvesting system is to obtain the best ratio of the area yielding run-off to either the area where run-off is being directed or the capacity of the storage structure (volume of water collected).

In this way, the water captured for crop production during run-off periods can be stored either directly in the soil for subsequent use by plants or in small farm reservoirs or collection tanks. This aids stabilization of crop production by enhancing soil moisture availability or allowing irrigation during a dry period within the rainy season or by extending crop production into the dry season.

Some factors to be considered regarding these run-off farming systems and reservoirs include: site selection, watershed size and condition, rainfall distribution and run-off, and water requirements of crops. Where a minimum water depth of about 1 m can be maintained in a reservoir, fish can be raised to provide additional food.

Fig. 4.15

Numerous water harvesting systems have been developed over the centuries, especially in arid areas. The principle of collecting run-off for crop production is also inherent to many other soil and water conservation technologies that apply the concept of run-off and runon areas at a microwatershed level, such as negarims, trapezoidal or "eyebrow" bunds and tied ridges.

BALANCED FERTILIZATION

Where the supply of nutrients in the soil is ample, crops are more likely to grow well and produce large amounts of biomass. Fertilizers are needed in those cases where nutrients in the soil are lacking and cannot produce healthy crops and sufficient biomass. Most soils in sub-Saharan Africa (SSA) are deficient in P. P is required not only for plant growth but also for N fixation. Unbalanced fertilization, for example mainly with N, may result in more weed competition, higher pest incidence and loss of quality of the product.

Unbalanced fertilization eventually leads to unhealthy plants. Therefore, fertilizers should be applied in sufficient quantities and in balanced proportions. The efficiency of fertilizer use will be high where the organic matter content of the soil is also high. In very poor or depleted soils, crops use fertilizer applications inefficiently. When soil organic matter levels are restored, fertilizer can help maintain the revolving fund of nutrients in the soil by increasing crop yields and, consequently, the amount of residues returned to the soil.

Cover Crops

Growing cover crops is one of the best practices for improving organic matter levels and, hence, soil quality.

The benefits of growing cover crops include:

- They prevent erosion by anchoring soil and lessening the impact of raindrops.
- They add plant material to the soil for organic matter replenishment.
- Some, *e.g.* rye, bind excess nutrients in the soil and prevent leaching.

- Some, especially leguminous species, *e.g.* hairy vetch, fix N in the soil for future use.
- Most provide habitat for beneficial insects and other organisms.
- They moderate soil temperatures and, hence, protect soil organisms.

A range of crops can be used as vegetative cover, *e.g.* grains, legumes and oil crops. All have the potential to provide great benefit to the soil. However, some crops emphasize certain benefits; a useful consideration when planning a rotation scheme. It is important to start the first years with (cover) crops that cover the surface with a large amount of residues that decompose slowly (because of the high C:N ratio). Grasses and cereals are most appropriate for this stage, also because of their intensive rooting system, which improves the soil structure rapidly.

In the following years, when soil health has begun to improve, legumes can be incorporated in the rotation. Leguminous crops enrich the soil with N and their residues decompose rapidly because of their low C:N ratio. Later, when the system is stabilized, it is possible to include cover crops with an economic function, *e.g.* livestock fodder. The selection of cover crops should depend on the presence of high levels of lignin and phenolic acids. These give the residues a higher resistance to decomposition and thus result in soil protection for a longer period and the production of more stable.

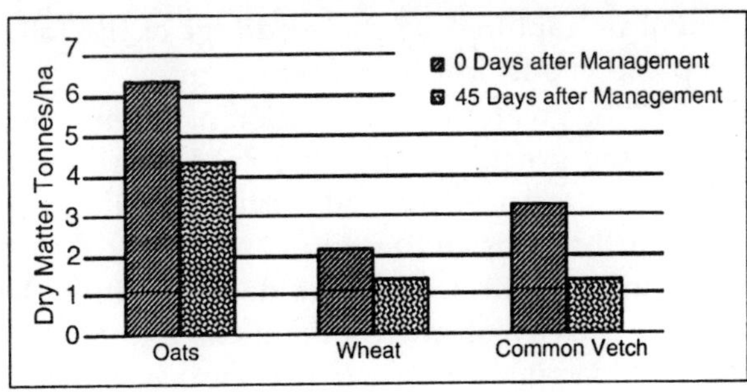

Fig. 4.16 Reduction of Dry Matter of Different Cover Crops

Another determining factor in the dynamics of residue composition is the biochemical composition of the residues. Depending on species, their chemical components and the time and way of managing them, there will be differences in decomposition.

The grain species (oats and wheat) show more resistance than common vetch (legume) to decomposition. The latter has a lower C:N ratio and a lower lignin content and is thus subject to a rapid decomposition.

Agricultural production systems in which residues are left on the soil surface, such as direct seeding and the use of cover crops, stimulate the development and activity of soil fauna at many levels.

The term green manure is often used to indicate the same plant species that are used as cover crops. However, green manure refers specifically to a crop in the rotation grown for incorporation of the non-decomposed vegetative matter in the soil. While this practice is used specifically to add organic matter, this is not the most effective use of organic matter (especially in hot climates) for two reasons:

1. Mechanical disturbance of the soil should be avoided as much as possible.
2. When biomass is incorporated in the soil all at one time, there is a short period of high microbial activity in decomposing the material. This results in the sudden release of a large quantity of nutrients that cannot be captured by the seedlings of the following crop and is thus lost from the system.

In general, the greater the production of green manure or crop biomass, the greater is the microbial, mesofauna and macrofauna population of the soil - from fungi and micro-organisms to earthworms and termites.

The dynamics of surface residue decomposition depend *inter alia*on the activity of micro-organisms and also on soil mesofauna and macrofauna. The macrofauna consists mainly of earthworms, beetles, termites, ants, millipedes, spiders, snails and slugs. These organisms help integrate the residues into the soil and improve soil structure, porosity, water infiltration, and through-flow through the creation of burrows, ingestion and secretions. The natural incorporation of cover-crop and weed residues from the soil surface to deeper layers in the soil by soil macrofauna is a slow process.

The activity of microorganisms is regulated by the activity of the macrofauna, because the latter provide them with food and air through their bioturbation activities. In this way, nutrients are released slowly and can provide the crop with nutrients over a longer period. At the same time, the soil is covered for a long time by the residues and is protected against the impact of rain and sun.

IMPROVED VEGETATIVE STANDS

In many places, low plant densities limit crop yields. Wide plant spacing is often practised as "a way to return power to the soil" or "to give the soil some rest", but in reality it is an indicator that the soil is impoverished. Plant spacing is usually determined by farmers in relation to soil fertility and available water or expected rainfall (unless standard recommendations are enforced by extension). This means that plants are often spaced widely on depleted soils in arid and semi-arid regions with a view to ensuring an adequate provision of plant nutrients and water for all plants.-

However, it is important to maintain the recommended plant spacing in order to optimize biomass production and rooting density and, hence, organic matter for food, moisture retention and habitat for soil organisms. Once the crop is established, reduced sunlight between closer crop rows may also reduce regrowth of weeds.

Planting Pits

Planting pits achieve fast rehabilitation of severely degraded land, especially in a semi-arid climate where a short fallow period of natural grass growth (2-6 years after 2-3 years under crops) cannot be expected to maintain or restore the land's agricultural productivity.An example of the rapid restoration of productivity of degraded land is an indigenous method in the Sahel region called "zaï". During the dry season, farmers dig out pits 15 cm deep and 40 cm in diameter every 80 cm, tossing the earth downhill.

The dry desert Harmattan wind blows various organic residues into the excavated pits. The organic materials are consumed quickly by termites, which excavate tunnels through the crusted surface, allowing the first rains to soak down deep, out of danger of direct evaporation. Two weeks before the onset of the rains, farmers spread one or two handfuls of dry dung (1- 2.5 tonnes/ha) in the bottom of the pits and cover it with earth to prevent the rains from eroding away the organic matter.

Millet is sown into the pits at the onset of the rainy season. As the first rains wash over the surface crust (of the degraded land), the basins capture this run-off (enough to soak a pocket of soil up to 1 m in depth). The sown seeds germinate, break up the slaked surface crust and send roots down to the deeper stores of both water and nutrients (recycled by the termites).

Fig. 4.17

At harvest time, stalks are cut at a height of 1 m and left *in situ* to reduce wind-speed and trap windborne organic matter. In the second year, the farmer either digs new basins between the first ones and dresses them with manure, or pulls up the stubble and sows again in the old basins. Stubble clumps laid between basins are in turn used as a food source by termites. Planting pits are a way of increasing biomass production and crop yields on severely degraded land in semi-arid conditions. Rainfall is concentrated near the plants, and soil faunal activity and organic matter accumulation are concentrated in the planting pits. Planting pits have been introduced successfully in Zambia as a conservation practice for smallholder farmers, who do not have fertilizers or tractor services available to them.

AGROFORESTRY AND ALLEY CROPPING

Agroforestry is a collective name for land-use systems where woody perennials (trees, shrubs, palms, etc.) are integrated in the farming system. Alley cropping is an agroforestry system in which crops are grown between rows of planted woody shrubs or trees. These are pruned during the cropping season to provide green manure and to minimize shading of crops.

Agroforestry covers a wide range of systems combining food crops, forestry and pasture species in different ways (agrosilviculture, silvipasture, agrosilvipasture and multipurpose forest production). There are two different approaches to agroforestry. One uses agricultural crops or pasture as a transitional means of utilizing the land until forest plantations are fully established. The other is to integrate trees and shrubs permanently into the crop or animal production system, to the benefit of both crop production and land resource protection. Thus, agroforestry encompasses many traditional land-use systems such as home gardens, shifting cultivation and bush fallow systems.

Examples of agroforestry systems worldwide:

- Poro (*Erythrina poeppigiana*) has been grown extensively in coffee plantations in Costa Rica for shade, soil enrichment, live mulching and live fences.
- *Albizzia* spp. have been used in tea plantations in many Asian countries.
- In Indonesia, leucaena (*Leucaena leucocephala*) has been planted as contour hedges on hillsides for erosion

control, soil improvement and green mulch. It is estimated that some 20 000 ha of undulating land have been converted to these systems.

- In West Africa and Rwanda, many farmers use trees, fruit trees, bushes and grasses planted with agricultural crops on their farms.
- Many coconut plantations in the Caribbean are partly planted with bananas or used as pastures. Some small farms in Jamaica plant coconuts, banana and citrus together.

Alley cropping can be considered an improved bush fallow system. Small trees or shrubs are planted in cropland in rows, preferably along the contour (even where east-west orientation of the rows may minimize shading of crops). The optimal spacing between rows depends on: slope; soil type and its susceptibility to erosion; rainfall; crop species; and the soil and crop management system.

Besides adding organic matter to the system, perennial trees and shrubs recycle plant nutrients from deeper soil layers through their rooting system. Through litter and pruning, these can be used again by annual crops. Probably the most important contribution of perennials in a production system lies in the fact that throughout the whole year their roots excrete root exudates and decaying root cells, which in turn are used as an energy source by soil microorganisms. The food web in the soil is maintained, even during dry seasons when no annual crops are grown. The result is that soil biota are in place to provide the crop with nutrients at the beginning of the next cropping season.

Direct seeding is the easiest and cheapest way of establishing hedgerows around fields or in the fields (alleys). However, emerging seedlings may not be able to compete with weeds without additional care. Therefore, starting plant growth in a nursery and transplanting may be necessary for some species. Other species may be established by cuttings. With good establishment, the plants will be better able to withstand both dry spells and browsing by livestock. Crucial to a successful establishment of the hedgerow is that the selected plants should be tall enough to outgrow the weeds at the time of the first crop harvest.

Fig. 4.18

During the cropping season, hedgerow pruning is needed in order to avoid shading of the crop. The timing, frequency and extent of pruning depend on the species used and the season. As a general rule, the lower the hedgerows and the taller the crop, the less frequently is pruning required. Fastgrowing plants such as *Leucaena leucocephala* and *Gliricidia sepium* may require pruning every six weeks during the cropping season. They are often pruned to a height of about 50 cm. Care must be exercised as too frequent pruning can result in tree dieback. The integration of trees and woody shrubs into the cropping system offers additional uses and many benefits, as mentioned by farmers using the Quezungual system in Honduras. However, farmers with short-term land tenure may not be interested in these benefits. Furthermore, the plantation of trees sometimes has an effect on the land tenure status; therefore tenants may not be allowed to establish trees on agricultural land. Agroforestry systems can also inhibit mechanization and may need increased labour inputs, especially for hedgerow pruning.

Farmers' Perceptions of the Quezungual System: Benefits and Disadvantages

The Quezungual system has many benefits according to local farmers:

- Improved soil moisture conservation, which permits

a good development of the crop even during the dry spells of 2-4 weeks halfway through the rainy season;

- Production of fuelwood and fruits from the trees and shrubs;
- Agricultural production is greater than in traditionally managed plots;
- Plots with the Quezungual system can be cultivated for longer periods than under the slash-andburn system;
- Timber trees can be cut after about 7 years and used for construction and/or sold;
- The mulch obtained through the pruning of the trees and shrubs protects the soil surface from the impact of rain showers, thus there is less soil erosion (even during the heavy rains produced by hurricane "Mitch" in 1998 there was little soil erosion);
- Minimal labour is required to establish and maintain the Quezungual system;
- The soil becomes more fertile and the effect of fertilizers on production improves;
- The workability of the soil improves because the soil becomes softer, hence less labour is needed during sowing;
- The Quezungual system provides shade for farmers while they work the plot;
- Harvested products, such as beans and maize, can be dried by hanging them over the tree trunks;
- Cattle can feed on the residues after maize and sorghum harvests;
- Mulch cover reduces the incidence of disease in the bean crop;
- The presence of trees and shrubs in the plot attracts animals and insects, *e.g.* Birds and butterflies.

Disadvantages mentioned by the farmers are:

- In the first year, the production is the same as or slightly less than grain production obtained with the traditional system;

- In the early years of implementation, the incidence of slugs in the bean crop is greater.
- Too much soil cover can impede seed germination;
- The shade of the Quezungual system can result in a higher incidence of disease during intense rainfall periods (because of greater humidity).

Fig. 4.19

The hedgerow species have to be selected carefully in order to avoid negative impacts on crop production because of the complex relationships (competition for light, water and nutrients, allelopathy, occurrence of pest and diseases, etc.) that are inherent to agroforestry systems.

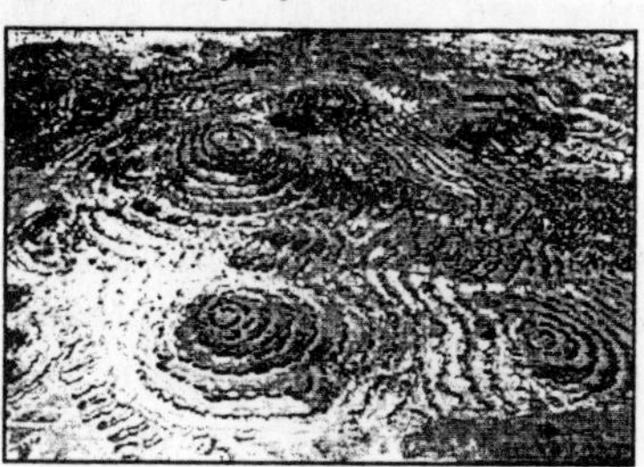

Fig. 4.20

Many farmers may consider the hedgerows as not useful, especially where their positive effects are not secure or visible. Where livestock are allowed to graze freely, it can be difficult to establish hedgerows without taking special measures to protect the young plants. Fencing or control of grazing animals may require collective efforts and agreement by the local community.

The increased labour requirement, the reduced cropped area and the difficulty of mechanization may make alley cropping uneconomic unless the hedgerow species produce direct benefits such as fruits, fuelwood or poles/timber for construction purposes (in addition to the nutrient recycling and erosion control effects).

REFORESTATION AND AFFORESTATION

Afforestation means the establishment of a forest on land that has not grown trees recently. It can serve two principal soil and water conservation purposes: protection of erosion-prone areas, and revegetation and rehabilitation of degraded land. Afforestation is specifically used to provide protective cover in vulnerable, steep and mountainous areas. Afforestation helps to replenish timber resources and provide fuelwood and fodder.

The establishment of a forest cover under good management is an effective means of increasing organic matter production. However, the land must have the productive capacity to support an appropriate forest type, which differs according to climate, soil, slope and the specific purpose of the forest (timber production, livestock grazing, etc.). Therefore, the choice of species and the selection of an appropriate site are of particular importance for successful afforestation.

The procurement of adequate quantities of good quality seed of the species and provenances (adapted varieties) required is a prerequisite for any afforestation effort. However, it is often difficult to find suitable and reliable sources of such seeds.

A number of species require special pre-treatment of the seed or seedling in order to achieve satisfactory germination and uniform stand. Such treatment may consist of soaking the seed in water for varying lengths of time, alternate soaking and drying, scarifying or chipping the seed coat to render it permeable to water, plunging the seed into boiling water or even boiling it for a short time. Some tree seedlings may need a mycorrhizal treatment when planted in soils that are deprived of associated mycorrhizae species as well as rhizobium species (*e.g.Casuarina* in Senegalese sandy soils). The aim is to ensure that good numbers of plants germinate and that germination after sowing is both rapid and uniform.

Afforestation can be achieved by direct sowing or replanting young plants from a nursery. The main advantage of direct sowing is the reduced cost.

However, this is usually much less reliable and is only justified where:

- Seed is plentiful and cheap;
- Adequate germination under field conditions can be relied on;
- The seedlings send down a deep tap root rapidly and are able to withstand adverse climatic conditions in the time after germination;
- The rate of growth is sufficiently fast to make a prolonged period of tending and weeding unnecessary.

Regeneration of Natural Vegetation

Regeneration of natural grasslands and forest areas increases biomass production and improves the plant species diversity, resulting in more diverse soil biota and other associated beneficial organisms. Natural regeneration may be more reliable where land is not very productive. In some cases, natural regeneration of a given area may lead to the infestation of plots by weeds. Increasingly, natural vegetation is being recognized for its multipurpose benefits, for example, fuelwood, fibre, biocontrol (*e.g.* neem) and medicinal species, as well as restoration of soil fertility (*Acacia albida* and other leguminous species) and habitats for various beneficial species (pollinators and natural enemies) as well as wildlife.

INCREASED ORGANIC MATTER SUPPLY

PROTECTION FROM FIRE

Burning affects organic matter recycling significantly. Fire destroys almost all organic materials on the land surface except for tree trunks and large branches. In addition, the surface soil is sterilized, loses part of its organic matter, the population of soil microfauna and macrofauna is reduced, and no ready-to-use organic matter is available for rapid restoration of the populations. However, this practice is widely used (*e.g.* in Africa) in order to enhance pasture regrowth for livestock (using residual P), to control pests and diseases, and even to catch small animals for food.

A specific and difficult case is the burning of sugar cane before harvesting. It has both a technical dimension (CO_2 and greenhouse gas emissions, mechanization of harvest, sugar content, etc.) and a social dimension (manual cutting, source of survival resources for poor/landless workers). The damage depends on fire intensity, which is a function of vegetation type and climate conditions and frequency. The costs and benefits of burning and the methods to minimize harmful effects need to be identified with local populations.

CROP RESIDUE MANAGEMENT

In systems where crop residues are managed well, they:

- Add soil organic matter, which improves the quality of the seedbed and increases the water infiltration and retention capacity of the soil, buffers the ph and facilitates the availability of nutrients;
- Sequester (store) C in the soil;
- Provide nutrients for soil biological activity and plant uptake;
- Capture the rainfall on the surface and thus increase infiltration and the soil moisture content;
- Provide a cover to protect the soil from being eroded;
- Reduce evaporation and avoid desiccation from the soil surface.

Depending on the nature of the following crop, decisions are made as to whether the residues should be distributed evenly over the field or left intact, *e.g.* where climbing cover crops (*e.g.* mucuna) use the maize stalks as a trellis.

An even distribution of residues:

- Provides homogenous temperature and humidity conditions at sowing time;
- Facilitates even sowing, germination and emergence;
- Minimizes the development of pests and diseases; and
- Reduces the emergence of weeds through allelopathic effects.

The most appropriate method for managing crop residues depends on the purpose of the crop residues and the experience and equipment available to the farmer. Where the aim is to maintain a mulch over the soil for as long as possible, the biomass is best managed using a knife roller, chain or sledge in order to break it down but not kill it. Where the decomposition process should commence immediately in order to release nutrients, the residues should be slashed or mown and some N applied because dry residues have a high C:N ratio. However, in order to avoid nitrate emission, urea should not be broadcast on the surface but injected where possible.

Fig. 4.21

UTILIZING FORAGE BY GRAZING RATHER THAN BY HARVESTING

In many places, there is competition for the use of crop residues that can be used as fodder, for roofing, artisan handicrafts, etc. Where residues are to be used for animal feed, either the animals graze the residues directly, or they are stall- or kraal-fed.

Removal of the residues from the field can lead to a considerable loss of organic matter where animal manure is not returned to the field. By controlled grazing, the animal manure is returned in the field without a high labour input.

The experience of Guaymango, El Salvador, demonstrates that it is possible to achieve successful integration of crop and livestock components without creating competition in the allocation of crop residues. The amount of residues produced by the system is enough to serve both as soil cover and as fodder for livestock, mainly because of the use of local sorghum varieties (instead of HYVs) that have a high straw/grain ratio. As farmers value crop residues as soil cover, a fodder market has developed where grazing rights, number of cattle and duration of grazing are traded.

In the northern zone of the United Republic of Tanzania, farmers have found a compromise between using the residues for grazing or soil cover, albeit one that is rather labour intensive. They separate the palatable and non-palatable parts of the crop residues. They use the non-palatable parts to cover the soil and act as food for soil organisms, while they feed the palatable parts to cattle and goats that are kept close to the homestead.

INTEGRATED PEST MANAGEMENT

As with balanced fertilization, proper pest and disease management results in healthy crops. Healthy crops produce optimal biomass, which is necessary for organic matter production in the soil. Diversified cropping and mixed crop-livestock systems enhance biological control of pests and diseases through species interactions. Through integrated production and pest management farmers learn how to maintain a healthy environment for their crops. They learn to examine their crops regularly in order to observe ratios of pests to natural enemies (beneficial predators) and cases of damage, and on that basis to make decisions as to whether it is necessary to use natural treatments (using local products such as neem or tobacco) or chemical treatments and the required applications.

Fig. 4.22

APPLYING ANIMAL MANURE OR OTHER CARBON-RICH WASTES

Any application of animal manure, slurry or other carbon-rich wastes, such as coffee-berry pulp, improves the organic matter content of the soil. In some cases, it is better to allow a period of decomposition before application to the field. Any addition of carbon-rich compounds immobilizes available N in the soil temporarily, as micro-organisms need both C and N for their growth and development. Animal manure is usually rich in N, so N immobilization is minimal. Where straw makes up part of the manure, a decomposition period avoids N immobilization in the field.

Fig. 4.23

COMPOST

Composting is a technology for recycling organic materials

in order to achieve enhanced agricultural production. Biological and chemical processes accelerate the rate of decomposition and transform organic materials into a more stable humus form for application to the soil. Composting proceeds under controlled conditions in compost heaps and pits.

Compost heaps should have a minimum size of 1 m^3 and are suitable for more humid environments where there is potential for watering the compost. Compost pits should be no deeper than 70 cm and should be underlain with rough material for good aeration of the compost. Pits are suitable for drier environments where the compost may desiccate. Dry composting relies on covering the compost with soil and creating an anaerobic environment. However, this is a slower process than the more usual moist aerobic process. The ratio of C to N in the compost pile is important for optimizing microbial activity. Thus, a mixture of soft, green and brown, tougher material is used. Ash and phosphate rock are often added to accelerate the process.

Composting can complement certain crop rotations and agroforestry systems. It can be used efficiently in planting pits and nurseries. It is very similar in composition to soil organic matter. It breaks down slowly in the soil and is very good at improving the physical condition of the soil (whereas manure and sludge may break down fairly quickly, releasing a flush of nutrients for plant growth). In many circumstances, it takes time to rejuvenate a poor soil using these practices because the amount of organic material being added is small relative to the mineral proportion of the soil.

Successful composting depends upon the sufficient availability of organic materials, water, manure and "cheap" labour. Where these inputs are guaranteed, composting can be an important method of sustainable and productive agriculture. It has ameliorative effects on soil fertility and physical, chemical and biological soil properties. Well-made compost contains all the nutrients needed by plants. It can be used to maintain and improve soil fertility as well as to regenerate degraded soil. However, materials for compost production may be in short supply and the technology demands high labour inputs for proper compost production and application. Therefore, compost application may be restricted to certain crops and limited application areas, *e.g.* vegetable production in home gardens.

MULCH OR PERMANENT SOIL COVER

One way to improve the condition of the soil is to mulch the area requiring amelioration. Mulches are materials placed on the soil surface to protect it against raindrop impact and erosion, and to enhance its fertility. Crop residue mulching is a system of maintaining a protective cover of vegetative residues such as straw, maize stalks, palm fronds and stubble on the soil surface.

The system is particularly valuable where a satisfactory plant cover cannot be established rapidly when erosion risk is greatest.

Mulching adds organic matter to the soil, reduces weed growth, and virtually eliminates erosion during the period when the ground is covered with mulch.

There are two principal mulching systems:

1. *In situ* mulching systems - plant residues remain where they fall on the ground;
2. Cut-and-carry mulching systems - plant residues are brought from elsewhere and used as mulch.

Crop residue mulching has numerous positive effects on crop production. However, it may require a change in existing cropping practices.

For example, farmers may conventionally burn crop residues instead of returning them to the soil. *In situ* mulching depends on the design of appropriate cropping systems and crop rotations, which have to be integrated with the farming system. The greater labour demands of cut-and-carry systems represent a major constraint. Mulch may be more relevant in home gardens or for valuable horticulture crops than in less intensive farming systems.

Mulch affects the soil life. Holland and Coleman have demonstrated that litter placement on the soil surface (as opposed to incorporation with ploughing) increased the ratio of fungi to bacteria - the reason being that fungi have a higher carbon assimilation efficiency than bacteria. In addition, it encourages bioturbating (mixing) effects of macrofauna that pull the materials into surface layers of the soil.

Fig. 4.24

DECREASED DECOMPOSITION RATES

REDUCED OR ZERO TILLAGE

Repetitive tillage degrades the soil structure and its potential to hold moisture, reduces the amount of organic matter in the soil, breaks up aggregates, and reduces the population of soil fauna such as earthworms that contribute to nutrient cycling and soil structure.

Avoiding mechanical soil disturbance implies growing crops without mechanical seedbed preparation or soil disturbance since the harvest of the previous crop. The term zero tillage is used for this practice synonymously with terms such as no-till farming, no tillage, direct drilling, and direct seeding.

Compared with conventional tillage, reduced or zero tillage has two advantages with respect to soil organic matter. Conventional tillage stimulates the heterotrophic microbiological activity through soil aeration, resulting in increased mineralization rate. Through breakdown of soil structure, it decreases upward and downward movements of soil fauna, such as earthworms, which are largely responsible for "humus" production through the ingestion of fresh residues. Reduced or zero tillage regulates heterotrophic microbiological activity because the pore atmosphere is richer in CO_2/O_2, and facilitates the activity of the "humifiers".

Fig. 4.25

Mulching in the highlands of northern Thailand:

- Why certain crops receive mulch and others do not
- Mulching provides a particular benefit to the cultivated crop. Mulching is practised for various cash crops for specific reasons. Onion and garlic are mulched mainly to control weeds (early hand weeding would be difficult without damaging the crop). The mulch is also important to keep the soil moist and cool as these crops are usually grown during the dry season under irrigation. Mulch is also applied under flowers and strawberries, mainly to protect the fragile and valuable products from becoming soiled.
- Mulching saves labour. Mulching is often seen in maize fields, before as well as after crop establishment. Maize can compete reasonably well with weeds. Therefore, some farmers plant maize without tillage in a mulch of weeds previously killed with herbicides - a system that is less labour-demanding than a tillage operation. Because maize is planted with large spacings, it generally requires less rigorous weeding.

Weeding is often done by slashing, and the weed residues are left on the ground.

Fig. 4.26

Tillage has become the most common method to control weeds. However, mulching is a more environmentally sound practice than tillage for weed control. The loose soil that results from tilling has less structure than before; the appearance is deceptive. Subsequent traffic or heavy rain soon packs this loosened soil, not only negating the expensive cultivation that produced the loose soil but also culminating in a degraded environment for water entry, seed germination and root growth. Further cultivation is then required to re-loosen the soil; more expense with the same outcome - subsequent repacking and degraded soil structure. This is a typical "downward spiral" of conventional agriculture. Moreover, tillage when the soil is too moist or too dry leads to compaction or pulverization of soil; but farmers may not have the option to wait for optimal conditions.

Severe, accelerated soil erosion and the high costs in terms of labour and energy associated with plough-based methods of seedbed preparation have led to the widespread adoption of no- or zero-tillage systems for cropping in temperate and tropical climates. In no-tillage systems, the crop is sown into a soil left undisturbed since the harvest of the previous crop. Crop residue mulch is maintained and anchored firmly to the ground. Weed control relies on mechanical slashing or cover crops. Contact herbicides are also used in some cases.

In reduced- or zero-tillage systems, soil fauna resume their bioturbating activities gradually. These loosen the soil and mix the soil components (also known as biotillage). The additional benefit of the increased soil organic matter and burrowing is the creation of a stable and porous soil structure without expensive, time-consuming and potentially degrading cultivations.

In zero-tillage systems, the action of soil macrofauna gradually incorporate cover crop and weed residues from the soil surface down into the soil. The activity of microorganisms is also regulated by the activity of the macrofauna, which provide them with food and air through their burrows. In this way, nutrients are released slowly and can provide the following crop with nutrients.

Several authors have demonstrated that some crop rotations and zero tillage favour *Bradyrhizobia* populations, nodulation and thus N fixation and yield. Figure 4.27 indicates a 200-300-per cent increase in population size of root nodule bacteria in a zero-tillage system compared with conventional tillage. The presence of soybean in the crop rotation resulted in a fivefold to tenfold increase in population size of the same bacteria compared with cropping systems without soybean.

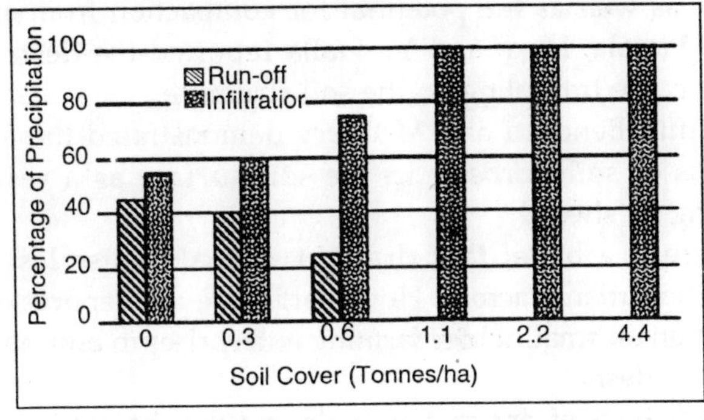

Fig. 4.27 Population Size of Root Nodule Bacteria with Different Crop Rotations

Note: S = soybean; W = wheat; M = maize.

Strictly speaking, the term zero tillage applies to methods involving no soil disturbance whatsoever, a condition that may be difficult to achieve. Broadcasting of seed is one way of applying zero tillage. The seed is broadcast over the previous crop residues and, where necessary, the residues are shaken to ensure that the seed falls on the soil surface.

In direct drilling, seeds such as maize, sorghum, soybean, wheat and barley are sown directly into shallow furrows cut into the previous crop residues. Weeds are controlled mechanically with a knife, which knocks down the plants and breaks their stems, or chemically with herbicides.

Traditional practices such as the burning of crop residues may inhibit the introduction of no-tillage systems. In many situations, a conflict exists between leaving crop residues on the surface or feeding them to livestock in the dry season when there is a shortage of fodder.

Mechanical soil disturbance also includes soil compaction through wheel impact of machinery, especially important in large-scale mechanized agriculture, *e.g.* plantations (sugar cane) or biannual crops (cotton).

In a zero-tillage farming system, consideration must be given to reducing both the random placement of tyres/wheels in fields as well as the potential for compaction from animal hooves. Pietola, Horn and Yli-Halla reported the destructive effect of cattle trampling on the soil structure.

Proffitt, Bendotti and McGarry demonstrated the almost total loss of soil porosity in the soil surface as a result of trampling by sheep.

There is a belief that draught animals cause less land degradation than tractors. However, there are reports of soil compaction on smallholder farming enterprises in both Malawi and Bangladesh.

The hooves of draught animals and the shearing effect of ploughs or hand hoes, which are used repeatedly at a constant depth, can cause severe compacted layers. Grazing animals should be removed from zero-till fields in moist-wet soil conditions as the compaction risk is greatest at these times.

Fig. 4.28

Controlled traffic, where the wheels of all in-field equipment follow permanent, defined tracks, ensures that compaction is restricted to specific known areas. Alternatively, flotation tyres (low ground-pressure tyres) should be fitted to all large tractors, harvesters, in-field grain bins, etc. in order to reduce their compacting potential. Recent research has demonstrated the devastating effects of compaction from wheel impact on the occurrence and survival of eartworms. Earthworm incidence was greater under controlled traffic than under wheeled traffic.

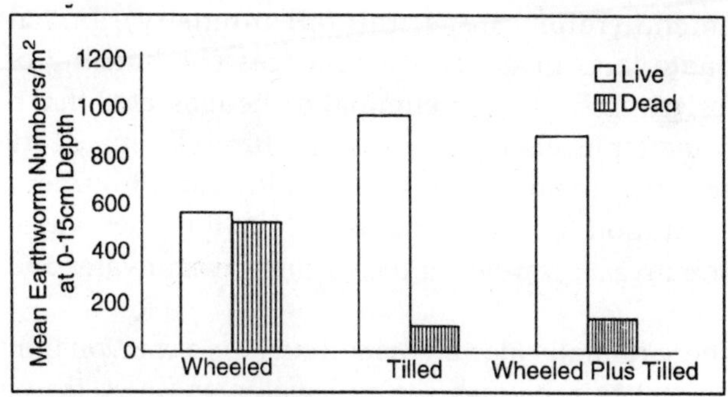

Fig. 4.29 Live and Dead Earthworm Numbers Per Square Metre at 0-15 cm of Soil Depth Sampled Immediately after Treatment

Figure 4.29 shows the immediate effects of wheeling and tillage on the earthworm population. It appears that wheeling has the most detrimental effect on earthworm survival and that where wheeling is followed by tillage the survival rate is much greater. This may indicate that earthworms are able to survive an initial compaction in the field as long as it is relieved immediately. Where it is not, earthworms are inmobilized and unable to find air and nutrients.

EFFECT OF SOIL ORGANIC MATTERON SOIL PROPERTIES

Organic matter affects both the chemical and physical properties of the soil and its overall health. Properties influenced by organic matter include: soil structure; moisture holding capacity; diversity and activity of soil organisms, both those that are beneficial and harmful to crop production; and nutrient availability. It also influences the effects of chemical amendments, fertilizers, pesticides and herbicides.

INEFFICIENT USE OF RAINWATER

Drylands may have low crop yields not only because rainfall is irregular or insufficient, but also because significant proportions of rainfall, up to 40 per cent, may disappear as run-off. This poor utilization of rainfall is partly the result of natural phenomena (relief, slope, rainfall intensity), but also of inadequate land management practices (*i.e.* burning of crop residues, excessive tillage, eliminating hedges, etc.) that reduce organic matter levels, destroy soil structure, eliminate beneficial soil fauna and do not favour water infiltration. However, water "lost" as run-off for one farmer is not lost for other water users downstream as it is used for recharging groundwater and river flows.

Where rainfall lands on the soil surface, a fraction infiltrates into the soil to replenish the soil water or flows through to recharge the groundwater. Another fraction may run off as overland flow and the remaining fraction evaporates back into the atmosphere directly from unprotected soil surfaces and from plant leaves.

The above-mentioned processes do not occur at the same moment, but some are instantaneous (run-off), taking place during a rainfall event, while others are continuous (evaporation and transpiration). To minimize the impact of drought, soil needs to capture the rainwater that falls on it, store as much of that water as possible for future plant use, and allow for plant roots to penetrate and proliferate. Problems with or constraints on one or several of these conditions cause soil moisture to be one of the main limiting factors for crop growth.

The capacity of soil to retain and release water depends on a broad range of factors such as soil texture, soil depth, soil architecture (physical structure including pores), organic matter content and biological activity. However, appropriate soil management can improve this capacity.

Practices that increase soil moisture content can be categorized in three groups:

1. Those that increase water infiltration;
2. Those that manage soil evaporation; and
3. Those that increase soil moisture storage capacities. All three are related to soil organic matter.

In order to create a drought-resistant soil, it is necessary to understand the most important factors influencing soil moisture.

INCREASED SOIL MOISTURE

Organic matter influences the physical conditions of a soil in several ways. Plant residues that cover the soil surface protect the soil from sealing and crusting by raindrop impact, thereby enhancing rainwater infiltration and reducing run-off. Surface infiltration depends on a number of factors including aggregation and stability, pore continuity and stability, the existence of cracks, and the soil surface condition. Increased organic matter contributes indirectly to soil porosity (via increased soil faunal activity).

Fresh organic matter stimulates the activity of macrofauna such as earthworms, which create burrows lined with the glue-like secretion from their bodies and are intermittently filled with worm cast material. The proportion of rainwater that infiltrates into the soil depends on the amount of soil cover provided (Figure 4.30). The figure 4.30 shows that on bare soils (cover = 0 tonnes/ha) run-off and thus soil erosion is greater than when the soil is protected with mulch. Crop residues left on the soil surface lead to improved soil aggregation and porosity, and an increase in the number of macropores, and thus to greater infiltration rates.

Increased levels of organic matter and associated soil fauna lead to greater pore space with the immediate result that water infiltrates more readily and can be held in the soil. The improved pore space is a consequence of the bioturbating activities of earthworms and other macro-organisms and channels left in the soil by decayed plant roots. On a site in southern Brazil, rainwater infiltration increased from 20 mm/h under conventional tillage to 45 mm/h under no tillage. Over a long period, improved organic matter promoted good soil structure and macroporosity. Water infiltrates easily, similar to forest soils (Figure 4.31).

The consequence of increased water infiltration combined with a higher organic matter content is increased soil storage of water. Organic matter contributes to the stability of soil aggregates and pores through the bonding or adhesion properties of organic materials, such as bacterial waste products, organic gels, fungal hyphae and worm secretions and casts. Moreover, organic matter intimately mixed with mineral soil materials has a considerable influence in increasing moisture holding capacity. Especially in the topsoil, where the organic matter content is greater, more water can be stored.

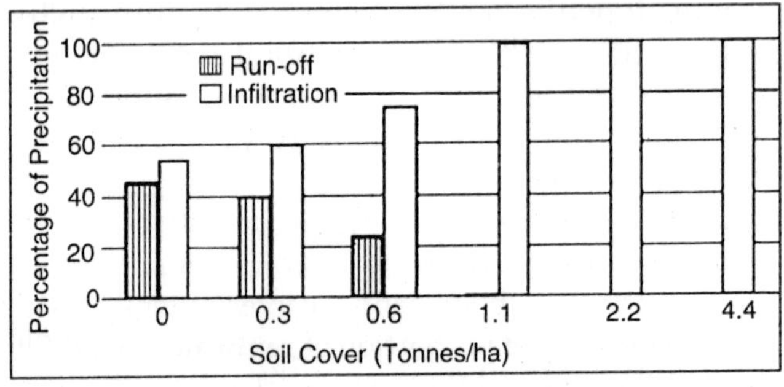

Fig. 4.30 Effect of Amount of Soil Cover on Rainwater Run-off and Infiltration

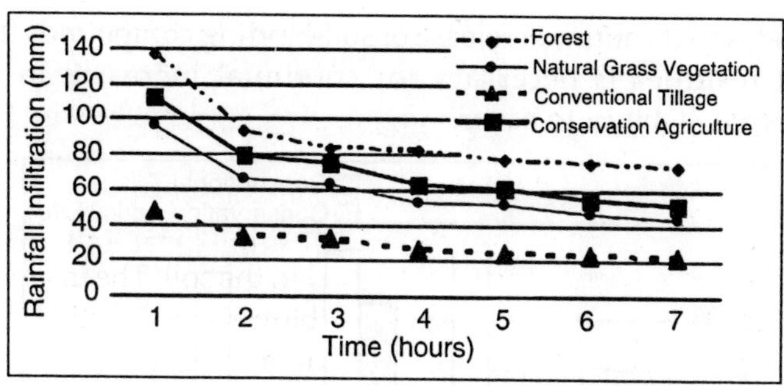

Fig. 4.31 Water Infiltration Under Different Types of Management

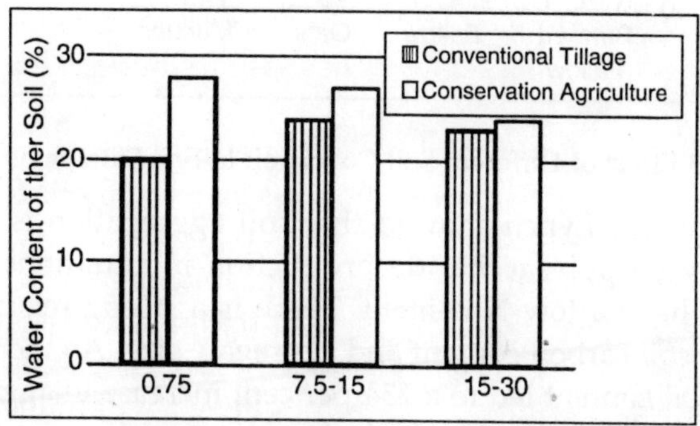

Fig. 4.32 Quantity of Water Stored in the Soil Under Conventional Tillage and Conservation Agriculture

The quality of the crop residues, in particular their chemical composition, determines the effect on soil structure and aggregation. Blair *et al.* report a rapid breakdown of medic (*Medicago truncatula*) and rice (*Oryza sativa*) straw residues resulting in a rapid increase in soil aggregate stability through the release of many soilbinding components.

As these compounds undergo further breakdown, they will be lost from the system resulting in a decline in soil aggregate stability over time.

The slow release of soil-binding agents from flemingia (*Flemingia macrophylla*) residues resulted in a slower but more sustained increase in the stability of soil aggregates. This indicates that continual release of soil-binding compounds from plant residues is necessary for continual increases in soil aggregate stability to occur.

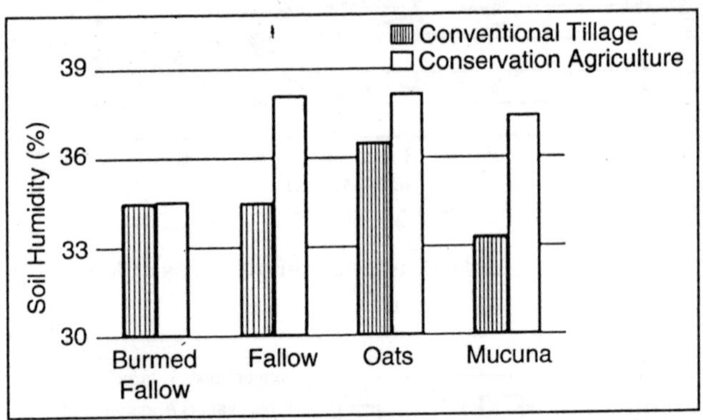

Fig. 4.33 Effect of Different Soil Covers on in-Soil Storage of Water

Elliot and Lynch showed that soil aggregation is caused primarily by polysaccharide production in situations where residues have a low N content. There is a strong relationship between soil carbon content and aggregate size. An increase in soil carbon content led to a 134-per cent increase in aggregates of more than 2 mm and a 38-per cent decrease in aggregates of less than 0.25 mm. The active fraction of soil C is the primary factor controlling aggregate breakdown.

In addition, although they do not live long and new ones replace them annually, the hyphae of actinomycetes and fungi play an important role in connecting soil particles. Gupta and Germida showed a reduction in soil macroaggregates correlated strongly with a decline in fungal hyphae after six years of continuous cultivation.

The in-soil storage of water depends not only on the type of land preparation but also on the type of cover or previous vegetation on the soil. Figure 4.33 indicates the effect of burning vegetation on the amount of water stored in the soil.

Conserving fallow vegetation as a cover on the soil surface, and thus reducing evaporation, results in 4 per cent more water in the soil. This is roughly equivalent to 8 mm of additional rainfall. This amount of extra water can make the difference between wilting and survival of a crop during temporary dry periods.

A study conducted in 1999 in Guatemala, Honduras and Nicaragua to evaluate the resilience of agro-ecosystems showed that 3-15 per cent more water was stored in the soil under more ecologically sound practices.

Unger showed that high wheat-residue levels resulted in increased storage of fallow precipitation, which subsequently produced higher sorghum grain yields. High residue levels of 8-12 tonnes/ha resulted in about 80-90 mm more stored soil water at planting and about 2.0 tonnes/ha more of sorghum grain yield compared to no residue management.

Table. Average Soil Depth at Which Moisture Starts, and Difference in Moisture Stored

Country	Agro-Ecologically Sound Practices cm	Conventional Practices cm	Difference (%)
Honduras	9.98	10.28	2.9
Guatemala	2.44	2.99	15.0
Nicaragua	15.81	17.80	11.2

The addition of organic matter to the soil usually increases the water holding capacity of the soil. This is because the addition of organic matter increases the number of micropores and macropores in the soil either by "gluing" soil particles together or by creating favourable living conditions for soil organisms.

Certain types of soil organic matter can hold up to 20 times their weight in water. Hudson showed that for each 1-per cent increase in soil organic matter, the available water holding capacity in the soil increased by 3.7 per cent. Soil water is held by adhesive and cohesive forces within the soil and an increase in pore space will lead to an increase in water holding capacity of the soil. As a consequence, less irrigation water is needed to irrigate the same crop.

Table. Economy of Irrigation Water Through Soil Cover, the Brazilian Cerrados

Country	Agro Ecologically Sound Practices (cm)	Conventional Practices	Difference (%)
Honduras	9.98	10.28	2.9
Guatemala	2.44	2.99	15.0
Nicaragua	15.81	17.80	11.2

REDUCED SOIL EROSION AND IMPROVED WATER QUALITY

The less the soil is covered with vegetation, mulches, crop residues, etc., the more the soil is exposed to the impact of raindrops. When a raindrop hits bare soil, the energy of the velocity detaches individual soil particles from soil clods. These particles can clog surface pores and form many thin, rather impermeable layers of sediment at the surface, referred to as surface crusts. They can range from a few millimetres to 1 cm or more; and they are usually made up of sandy or silty particles. These surface crusts hinder the passage of rainwater into the profile, with the consequence that run-off increases. This breaking down of soil aggregates by raindrops into smaller particles depends on the stability of the aggregates, which largely depends on the organic matter content.

Increased soil cover can result in reduced soil erosion rates close to the regeneration rate of the soil or even lower, as reported by Debarba and Amado for an oats and vetch/maize cropping system (Figure 4.34).

Soil erosion fills surface water reservoirs with sediment, reducing their water storage capacity. Sedimentation also reduces the buffering and filtering capacity of wetlands and the flood-control capacity of floodplains. Sediment in surface water increases wear and tear in hydroelectric installations and pumps, resulting in greater maintenance costs and more frequent replacement of turbines. Sediments can also reach the sea, harming fish, shellfish and coral. Eroded soil contains fertilizers, pesticides and herbicides; all sources of potentially harmful off-site impacts.

When the soil is protected with mulch, more water infiltrates into the soil rather than running off the surface. This causes streams to be fed more by subsurface flow rather than by surface run-off. The consequence is that the surface water is cleaner and resembles groundwater more closely compared with areas where erosion and run-off predominate. Greater infiltration should reduce flooding by increased water storage in soil and slow release to streams. Increased infiltration also improves groundwater recharge, thus increasing well supplies.

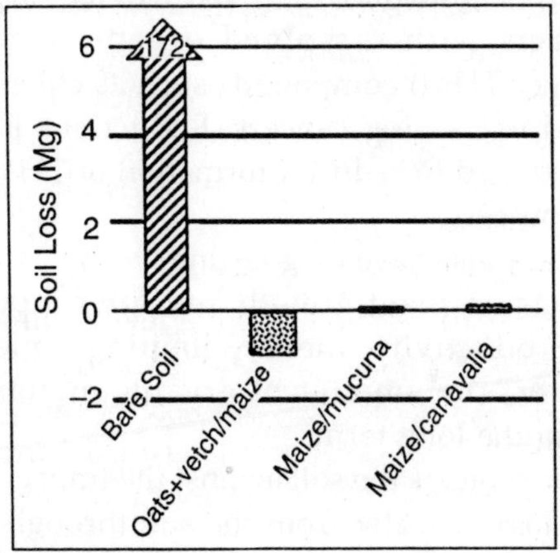

Fig. 4.34 Soil Loss Due to Water Erosion for Different Maize Cropping Systems

Bassi reported significant reductions in water turbidity and sediment concentration over a period of ten years in different catchment areas in southern Brazil. The reductions varied between 50 and 80 per cent depending on locally predominant soil types. These reductions were caused by increases in the incidence of planting perennial crops (banana and pasture) on hillsides, thereby decreasing erosion potential. Total sediment loss decreased by 16 per cent and the cost of fertilizers declined by 21 per cent; an indication of the previous loss of fertilizers with the eroded soil. Guimarães, Buaski and Masquieto illustrate the same effect for one specific catchment. The catchment area of Rio do Campo, Paraná, provides 80 per cent of the water supply for Campo Mourão, a city with an urban population of 357 000. In the period 1982-1999, a drastic reduction in water turbidity was measured.

Sediment and dissolved organic matter in surface water have to be removed from drinking-water supplies. Reduced erosion, and hence fewer soil particles in suspension, lead to lower costs for water treatment. Data from Chapecó, Brazil, indicate that the quantity of aluminium sulphate used for flocculating suspended solids fell by 46 per cent in five years. Where water is chlorinated to kill disease organisms, the chlorine reacts with dissolved organic matter to form trihalomethane (THM) compounds such as chloroform. THMs are suspected of causing cancers. Reductions in run-off and erosion should lead to reduced formation of THMs during the chlorination process.

Erosion may also have long-lasting secondary consequences through effects on plant growth and litter input. If erosion suppresses productivity, thereby limiting replenishment of organic matter, the amount of organic matter may spiral downwards in the long term.

Soil cover protects the soil against the impact of raindrops, prevents the loss of water from the soil through evaporation, and also protects the soil from the heating effect of the sun. Soil temperature influences the absorption of water and nutrients by plants, seed germination and root development, as well as soil microbial activity and crusting and hardening of the soil.

Roots absorb more water at higher soil temperatures up to a maximum of 35 °C. Higher temperatures restrict water absorption. Soil temperatures that are too high are a major constraint on crop production in many parts of the tropics. Maximum temperatures exceeding 40 °C at 5 cm depth and 50 °C at 1 cm depth are commonly observed in tilled soil during the growing season, sometimes with extremes of up to 70 °C. Such high temperatures have an adverse effect not only on seedling establishment and crop growth but also on the growth and development of the micro-organism population. The ideal rootzone temperature for germination and seedling growth ranges from 25 to 35 °C. Experiments have shown that temperatures exceeding 35 °C reduce the development of maize seedlings drastically and that temperatures exceeding 40 °C can reduce germination of soybean seed to almost nil.

Mulching with crop residues or cover crops regulates soil temperature. The soil cover reflects a large part of solar energy back into the atmosphere, and thus reduces the temperature of the soil surface. This results in a lower maximum soil temperature in mulched compared with unmulched soil (Figure 4.35) and in reduced fluctuations.

Fig. 4.35 Development of Water Turbidity Rates

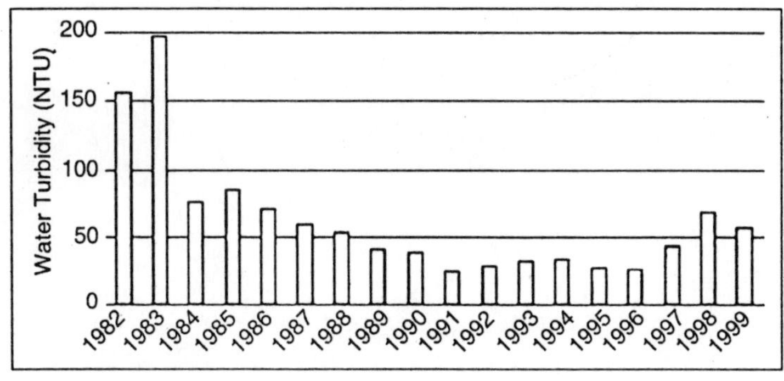

Fig. 4.36 Temperature Fluctuations at a Soil Depth of 3 cm in a Cotton Crop with and Without a Soil Cover of Mucuna

With Cover
Without Cover
Soil Temperataure °C
60
50
40
30
20
1 3 5 7 9 11 13 15 17 19 21 23
Time of the Day (Hours)

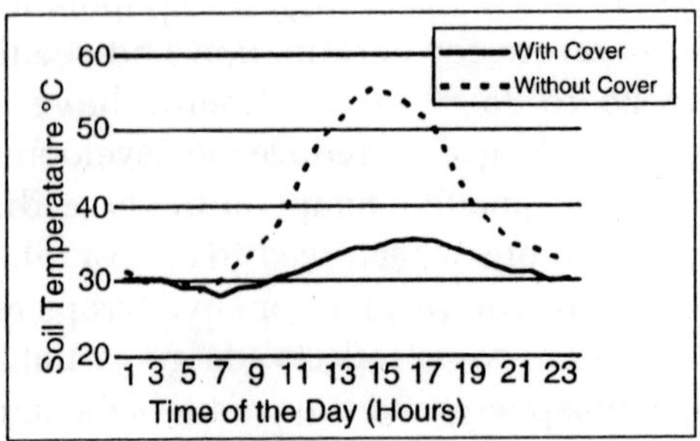

Fig. 4.37

5

Atmospheric Nanoparticles

INTRODUCTION

Atmospheric nanoparticles, defined as particles with spherical equivalent diameters smaller than 50 nm, are either directly emitted from combustion sources, or are formed in the atmosphere by a process called nucleation. They then quickly grow by the condensation of gas monomers or clusters, or by coagulation with other particles, to become a critical participant in a number of important atmospheric processes such as heterogeneous chemistry, cloud formation, precipitation, and the scatter and absorption of solar radiation. An understanding of nanoparticles has increased dramatically in recent years because of significant advances in instrumentation to detect, size, and determine their chemical composition. Recent observations now show us that atmospheric nano-particles are ubiquitous: Measurements carried out in city centers, isolated islands and forests, and the remote troposphere have never failed to encounter periods characterized by concentrations of up to 106 nanoparticles/cm3. Theoretical advances in nucleation and growth, at present, have not made the same sort of progress as observation, but this is an area of rapid progress.

The goal of this article is to provide an overview of the role that nanoparticles play in the atmosphere, the observational and theoretical tools currently being employed in their study, and recent observations that continue to direct progress in understanding their formation and fate. Although much is known about these transient particles, researchers appear to be at the cusp of huge scientific advances that will make it possible to predict their formation and behaviour in the atmosphere.

BACKGROUND

Excellent texts are available on the subject of atmospheric particulate matter, which is commonly referred to as atmospheric aerosol. The goal of this section is to provide a broad overview of the distribution, sources, and lifetimes of atmospheric aerosol. This is followed by a review of relevant atmospheric process involving nanoparticles.

ATMOSPHERIC PARTICLE SIZE DISTRIBUTIONS

Close to the Earth's surface, continental air typically contains from 103 to 105 particles/cm^3. These particles have spherical equivalent diameters dp ranging from 1 nm to 100 mm. Plots of particle number concentration, surface area, and volume vs. dp, such as those shown in Fig. 5.1 for typical rural settings, usually shows that these particles are distributed over three or more modes. Most particles are smaller than 100 nm in diameter, in a mode referred to as the Aitken mode.

The term ultrafine aerosol is often used for those with diameters smaller than 100 nm. The other two predominant modes are the accumulation mode (0.1 <dp<2.5 mm) and the coarse mode (dp>2.5 mm). Within the Aitken mode, for dp< 50 nm, many properties of the condensed phase, most notably vapour pressure, become quite different from those of the bulk phase: Thus particles smaller than 50 nm have come to be known as nanoparticles.

The implications of the distributions in Fig. 5.1 on atmospheric processes can be generalized as follows: For phenomena that depend on the particle number concentration, it is the Aitken mode that contributes the most; for those processes that are surface area dependent, the accumulation mode is most influential; for those that depend on total aerosol volume or mass, the coarse mode dominates.

ATMOSPHERIC PARTICLE SOURCES AND SINKS

The different modes in aerosol size distribution reflect differences in the atmospheric sources, transformations, and sinks (Fig. 5.1). Particles in the coarse mode are generated by mechanical processes such as wind and friction. The accumulation mode contains particles that have grown in the atmosphere by condensation of gases, or coagulation with other particles. The Aitken mode contains particles that are generated chemically, either directly emitted from a source (primary aerosol) or are created or grown from gas phase reactions in the atmosphere (secondary aerosol). The dominant loss mechanisms for Aitken mode particles are coagulation and condensational growth. Particles in the accumulation mode leave the atmosphere by adhering onto wetted surfaces, most commonly rain and cloud drops, through a process called wet deposition.

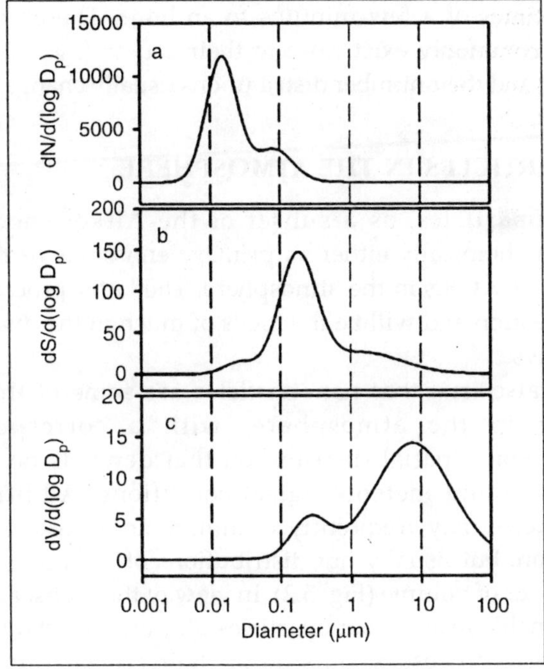

Fig. 5.1 Particle

Note:

a. Number,
b. Surface area, and
c. Volume distributions for typical rural conditions, generated using the parameterization of Jaenicke.

Coarse mode particles eventually settle to the ground in a process called dry deposition. Fig. 5.1 shows the typical timescales for these removal processes. Dry deposition, coagulation, and condensational growth are relatively rapid processes, whereas wet deposition is the least efficient means of particle removal. The result is that particles in the accumulation mode are the longest-lived particles, having lifetimes that may span several days.

Therefore these particles are commonly spatially homogeneous, and number distributions in this size range do not change rapidly with time. In contrast, Aitken mode particles have lifetimes of a few minutes to an hour. Therefore these particles commonly exist close to their source (*e.g.*, near busy freeways) and their number distributions usually change rapidly with time.

NANOPARTICLES IN THE ATMOSPHERE

Nanoparticles, as a subset of the Aitken mode, are generated chemically either as primary emissions or through gas phase reactions in the atmosphere. The latter process is, in fact, nucleation and will be the focus of much of the discussion that follows.

It is also true that nanoparticles are some of the most transient in the atmosphere, with a corresponding heterogeneous spatial distribution that depends on source distribution and meteorological conditions. Additionally nanoparticles may frequently dominate an aerosol number distribution, but usually not distributions of surface area and almost never of volume (Fig. 5.2). In view of these observations, it is reasonable to ask if nanoparticles play an important role in atmospheric processes, outside of their crucial role as the primary source of new particles. A question of equal importance, although outside the scope of this article, is whether atmospheric nanoparticles play a deleterious role in human health through respiration. The answer to this question is a qualitative "yes," although the mechanism and magnitude of this effect are far from clear. The contribution of nanoparticles to scattering and absorption of solar radiation, which controls processes such as visibility reduction and climate, is understood to be negligible. This is because of the low efficiency with which nanoparticles interact with solar radiation.

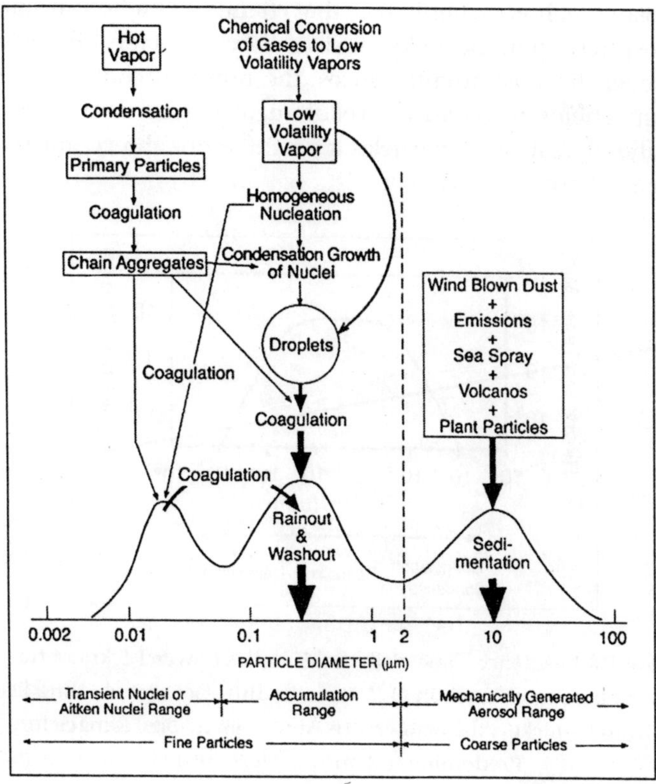

Fig. 5.2 Schematic of the Distribution of Particle Surface Area in the Atmosphere. Modes, Sources, and Sinks are Indicated.

Perhaps the most important contribution of nanoparticles on climate is to influence the optical properties, abundance, and persistence of clouds. An abundance of new particles could increase the number of cloud condensation nuclei (CCN), which are water-soluble aerosols on which water vapour condenses to form cloud droplets.

The overall result of this so-called "indirect effect of aerosols on climate" is an issue of active research and debate, but is generally thought to be a reduction in incoming solar radiation. At the time of this writing, very little is known about whether nanoparticles may play a unique role in the chemistry of the atmosphere. It is certainly true that chemical reactions occur on atmospheric nanoparticles, just as they occur on all aerosol; however, in most circumstances, the minimal contribution of nanoparticles to overall aerosol surface area decreases the likelihood that these particles may influence the chemistry of the gas phase.

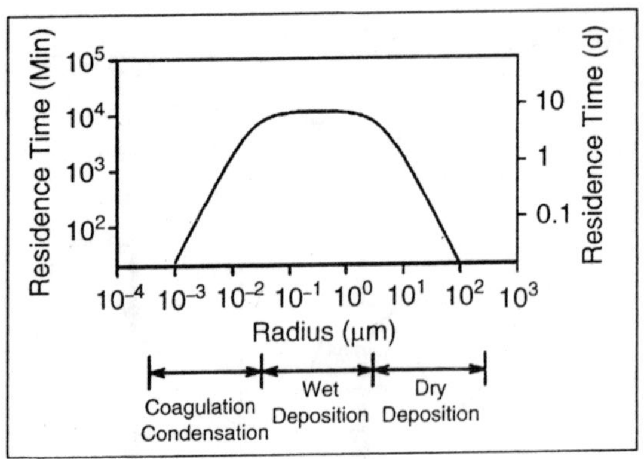

Fig.5.3 Residence Time of Particles in the Lower 1.5 km of the Atmosphere as a Function of Particle Radius, Generated Using Data from Jaenicke. Also Shown are Approximate Size Ranges for Predominant Removal Mechanisms.

The exception to this may be during "nucleation bursts," which are periods of intense new particle production that can dominate the surface area distribution.1-12-1 Although it may be true that the concentrations of certain reactive compounds may be enhanced in nanoparticles through the processes of nucleation and condensational growth, a simple analysis has shown that, even during nucleation burst periods, this enhancement would need to be a factor of 106 greater than reactant concentrations in typical cloud droplets to have a similar effect.

OBSERVATIONS OF ATMOSPHERIC NANOPARTICLE PHYSICOCHEMICAL PROPERTIES

Most of the important advances in our understanding of atmospheric nanoparticles have resulted from dramatic improvements in instrumentation for characterizing their size, number, and composition. As we will see later, these observations have challenged our theoretical understanding of both aerosol nucleation and growth. This section will provide an overview of these instruments, and some representative measurements.

Physical Characterization

Advances in the physical characterization of atmospheric nanoparticles now allow us to size and detect them when they are less than an hour old, corresponding to diameters of about 3 nm. It is now possible to characterize nanoparticle growth rates; thus we can start to make links between the presence of these nanoparticles and the variety of phenomena that they control. The current section briefly summarizes these physical characterization techniques.

The condensation nucleus counter (CNC) can detect particles as small as 3 nm in diameter. It accomplishes this by flowing the sample air first through a region that is saturated with a condensing vapour, usually n-butanol, and then through a cooled region. The gas then becomes supersaturated with the vapour, which causes the particles to grow to a size that can be detected by light scattering. The minimum detectable particle size is a function of vapour supersaturation; thus two CNCs with different supersaturations (or different transmission characteristics) can be operated in parallel and, by subtracting their readings, can be used to determine the concentration of particles within a certain size range. This subtraction technique is the most frequently used way of observing atmospheric nanoparticles. Recent advances have been reported in the development of particle size mag-nifiers, which are similar to CNCs but are being developed with the goal of detecting particles at diameters of 1 nm and below.

Size classification of nanoparticles is now possible using a number of techniques. The differential mobility analyser (DMA) is an "electrical mobility classifier" device that can size-classify particles down to 3 nm in diameter. It does this by charging particles and passing them through a region where a sheath gas and an electric field are combined in such a way that only particles of a certain surface area-to-charge ratio can pass through the exit aperture.

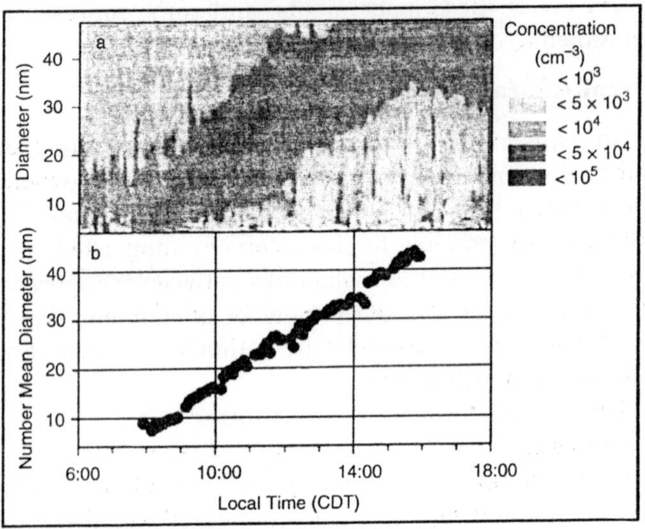

Fig. 5.4 (a) Particle Size Distribution vs. Local Time for a Nucleation Event. (b) Number Mean Diameter vs. Local Time for Distributions in (a), Showing a Linear Nanoparticle Growth Rate.

These instruments are commonly combined with CNCs to obtain nanoparticle size distributions. Fig. 5.4 a shows an example of a continuous record of particle size distributions from an urban site, which features a new particle formation event that reached detectable levels at 8:00 a.m. Subsequent growth of the mean diameter of the spectrum is quite linear (Fig. 5.4 b). This can be equated with aerosol growth if one assumes that the aerosol was homogeneous in a large-scale air mass.

Two additional methods for obtaining nanoparticle size distributions are the pulse height analysis (PHA) and the ion mobility spectrometer (IMS) techniques. PHA operates on the principle that particles smaller than 10 nm grow by condensation in the CNC to reach a unique final diameter, which can be sized by optical techniques. The IMS can measure the size spectra of charged nano-particles down to diameters below 1 nm, by measuring their drift velocity in a constant electric field. IMS measurements have shown that, under certain conditions, nanoparticle formation is correlated with bursts of atmospheric ion clusters.

Chemical Composition

Until recently, the chemical characterization of atmospheric nanoparticles has not seen the same sort of progress as physical characterization.

However, currently, several techniques have been developed for obtaining chemical composition information by:

- Indirectly inferring it by measuring some other behaviour;
- Using off-line collection and analysis techniques;
- Using online real-time mass spectrometry-based techniques. In this section, these various methods will be briefly described.

Some of the earliest measurements of the chemical properties of nanoparticles were performed by using a DMA to size-select particles, then exposing these particles to a controlled environment and acquiring a particle size distribution on the result to observe any changes in aerosol size. Most of these instruments, called tandem differential mobility analysers (TDMAs), expose size-selected particles to high humidity to investigate their hygroscopic properties, but others characterize volatility by applying high temperatures or organic composition by using a gas saturated with various organic vapors. In a related application, the growth characteristics of aerosol in a CNC have been investigated using the PHA technique, which was used to infer that nanoparticles in a boreal forest region were composed primarily of organic acids.

Off-line sampling techniques involve collecting particles on a substrate, often in a size-segregated manner, for later chemical analysis. Such analysis might involve extraction of the constituents from the substrate and analysis of the integrated composition by techniques such as ion chromatography and inductively coupled plasma mass spectrometry (ICP-MS), illuminating the substrate with a light source to investigate optical absorption, or the study of isolated particles on the substrate by microan-alytical techniques such as secondary ion mass spectros-copy (SIMS) and Fourier transform infrared spectroscopy (FTIR).

The application of off-line techniques to nanoparticle composition is challenged by an inability to collect sufficient mass of these particles; nonetheless, most of what is known about the composition of the smallest particles in the atmosphere is derived from offline techniques. Low-pressure impactors are often employed for collecting size-segregated ultrafine parti-cles, but these devices do not usually extend into the sub-50 nm diameter range.

Observations of ultrafine urban aerosol from low-pressure impactors show particles to be composed of 50-70% (by weight) organic compounds and 6-14% (by weight) of each of the following: elemental carbon, sulfate, nitrate, and trace metals. Recently, a particle concentrator that allows the collection of particles as small as 10 nm in diameter, requiring integration times of 3 hr or more, has been described. That study found a distinct mode in the 36-50-nm diameter range, affected primarily by combustion processes; however, correlations between elemental and organic carbon compositions also suggested that these particles contained secondary organic compounds. Dimethyla-mine has been identified by the chemical analysis of newly formed particles collected in a low-pressure im-pactor in a boreal forest. One recent example of the use of off-line microanalytical techniques has been reported in the study of recently formed particles in a coastal setting. That study observed both iodine and sulfur in particles with diameters below 10 nm, suggesting that biogenic iodine species emitted from seaweeds may be responsible for new particle formation or growth.

On-line chemical analysis techniques directly analyse particles in real time, usually by vaporizing particles through lasers or heated surfaces, ionizing the resulting gas, and injecting the ions into a mass spectrometer for analysis. These techniques are highly desirable for the study of nanoparticle composition, as they have capabilities of both high sensitivity and short sampling times. Most current on-line techniques rely on improving the nanoparticle sampling efficiency of existing instruments.

This has been accomplished primarily by the use of aerodynamic focusing lens sys-tems whose transmission efficiency degrades significantly for particle diameters smaller than 20 nm. Most of these instruments are very new—field measurements have only just started; therefore very little published data are available. One example of published results comes from the 1999 Atlanta Southern Oxidant Study, which found that particles as small as 14 nm are almost completely organic. The authors point out the ability to analyse nanoparticles is based, to some degree, on particle composition. For example, sulfate was not identified in any particles with their instrument, whereas other measurements during the same field campaign identified sulfate as a major constituent.

New instruments that have been designed specifically for the on-line characterization of nanoparticles have also been introduced. Most of these new devices are limited to obtaining atomic, rather than molecular, composition of the aerosol, and usually employ high-power lasers and plasmas to desorb and ionize aerosols smaller than 20 nm in diameter. The author has recently reported the development of the thermal desorption chemical ioniza-tion mass spectrometer (TDCIMS), an instrument capable of on-line measurements of the molecular composition of nanoparticles as small as 5 nm in diameter at time resolutions of ca. 20 min.

The TDCIMS operates by charging and then collecting nanoparticles on a metal filament, then resistively heating the filament and analyzing the desorbed gas by chemical ionization mass spectrometry (CIMS). Fig. 5.5shows an example of continuous TDCIMS measurements of sub-20-nm-diam-eter aerosols performed outside our laboratories in Boulder, CO, in the spring of 2002.

The major ion observed is ammonium; however, the integrated concentration of ions larger than 65 amu varies greatly during the day. The TDCIMS has also performed measurements of the composition of freshly nucleated particles at the 2002 Aerosol Nucleation and Real-time Characterization Experiment (ANARChE) in Atlanta, GA. Preliminary results from that experiment suggest that freshly nucleated aerosols in Atlanta are composed almost entirely of sulfate with variable degrees of neutralization by ammonium.

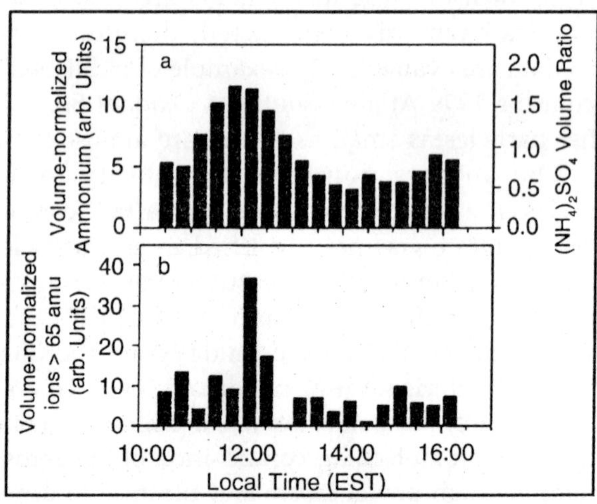

Fig. 5.5 Measurements of the Chemical Composition of Sub-20 nm Diameter Aerosol.

a. Ammonium ion concentration, normalized by collected aerosol volume and as the ratio to the ion signal from an equivalent volume of ammonium sulfate aerosol.
b. Integrated volume-normalized concentration of ions with molecular weight greater than 65 amu.

ATMOSPHERIC NANOPARTICLE FORMATION AND GROWTH: MODELS AND OBSERVATIONS

Given the current understanding of the atmospheric impact of nanoparticles, it is clear that the ability to predict the physicochemical properties of nanoparticles requires equal effort placed on understanding the growth of nanoparticles and their formation.

Because instruments are, thus far, unable to directly observe newly formed aerosol in the atmosphere, the interpretation of observations of nucleation also requires an understanding of the mechanism and dynamics of condensational growth. The goal of the present section is to review the current theoretical basis for nucleation and growth.

Figure 5.6 shows a schematic of the formation and growth of atmospheric nanoparticles. The primary formation process is homogeneous nucleation, which is defined as the formation of thermodynamically stable particles from the condensation of gaseous precursors. Primary sources, such as diesel engines, can be significant sources of nanoparticles in urban settings but will not be discussed in this article. Heterogeneous nucleation, defined as condensation of gases on foreign media such as gas phase ions, might also lead to the formation of atmospheric nanoparticles.

It is important to note that, in the atmosphere, homogeneous nucleation is in competition with scavenging of the low-volatility gas by preexisting aerosol. Because of this, new particle formation in the atmosphere tends to occur in bursts (Fig. 5.6), either when the total aerosol surface area is suddenly dropped, or when a sudden meteorological or chemical change modifies the concentration of condensable vapour.

CONDENSATIONAL GROWTH

One of the most unique properties of nanoparticles involves their evaporation and condensation behaviour. As the curvature of a pàrticle surface decreases, the separation between adjacent surface molecules increases leading to an overall decrease in the attractive forces between them. The result of this is that, at equilibrium, the partial pressure surrounding the curved surface will exceed the saturation vapour pressure for the flat surface. The equation that describes this effect is known as the Kelvin, or Thomson-Gibbs, equation:

$$S = \frac{Pd}{Ps} = \exp\left(\frac{4\sigma Mw}{\rho RTd*}\right)$$

Here the vapour saturation ratio S is defined as the ratio of the vapour pressure surrounding the particle pd to the saturation vapour pressure of the flat surface ps.

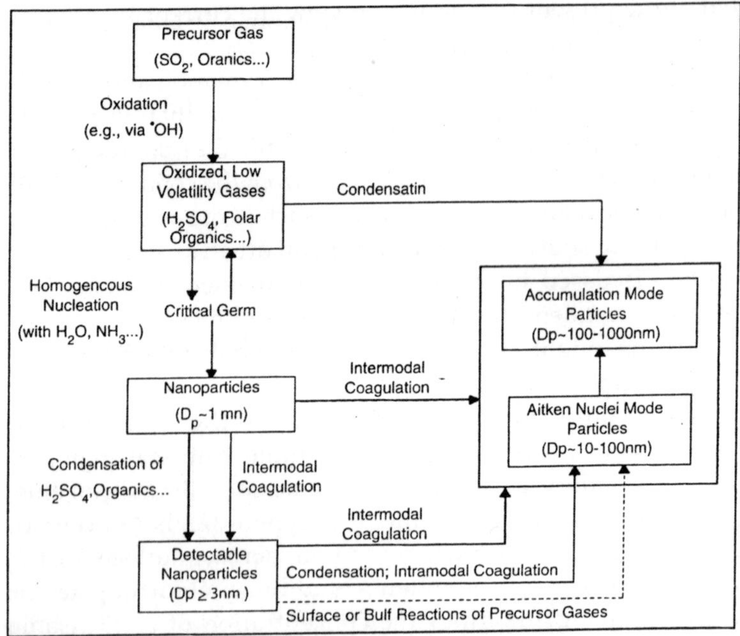

Fig. 5.6

The surface tension, molecular weight, and density of the liquid are given by s, Mw, and p, respectively. R is the universal gas constant, T is temperature, and d* is the diameter of a particle that will neither grow nor evaporate at S. Fig. 5.6 shows the relationship between S and d*.

The line given by Eq. can be thought of as the boundary between condensational growth and evaporation. If a particle's diameter and saturation ratio place it on the line and S were suddenly lowered, then the droplet will completely evaporate; if, conversely, S were raised, then the particle will grow indefinitely. The Kelvin equation predicts that a pure 10-nm diameter water particle would require a minimum relative humidity of 125% to prevent evaporation.

SOURCES OF ATMOSPHERIC NANOPARTICLES

The detection of nanoparticles and the measurement of specific properties associated with them are necessary for two distinctly different reasons.

First, methods are required that reliably detect nanoparticles and measure their physico-chemical properties in the media in which humans and ecosystems are exposed to them, such as air, water, soil, consumer products and nanocomposites. Methods must also be available that support the studies to assess the risk of nanoparticles, such as with toxicological and ecotoxicological studies. Additional tools are required in this case, which are able to detect nanoparticles in the relevant medium, including cells, fluids and plant tissue.

The second issue relates to the physical or chemical properties associated with nanoparticles that are the basis of the detection of nanoparticles in these media. The range of properties of nanoparticles of potential relevance to risk assessment highlights the principal needs for extremely sensitive methods. The typical dimensions of nanoparticles are below the diffraction limit of visible light, so that they are out of range for optical microscopy. In low concentration liquids and gases however, single chromophore detection is possible and in Scanning Near Field Optical Microscopy (SNOM) subwavelength feature can be analysed. While the chemical composition of nanoparticles might be accessible by classic analytic methods for macroscopic amounts of nanoparticulate material, chemical analysis of individual nanoparticles in a dilute environment was for a long time impossible due to their low mass, and only recently have methods become available for this purpose, so that even surface coatings may be detected.

NANOPARTICLES IN GAS SUSPENSION

Driven by the recent developments in atmospheric chemistry and physics, which have highlighted the role of nanoparticles in areas such as climate and health research, the measurement technologies for atmospheric aerosols offers a suite of tools specialized to the nanometer size range. An HSE Report and Luther have summarized the various types of device which might be or have been used to provide measurement information on nanometre size aerosols. Due to the lack of significant scattering or absorption by particles in the nanometre size range, particles are counted in commercially available, condensation nucleus counters (CPC or CNC), in which the particles are activated to droplets in a supersaturated atmosphere of alcohol, which can then be detected optically. Currently available instruments can detect particles as small as 3 nm, while new developments may reach the 1 nm limit. CPCs cover a large dynamic range from a few up to 10^6particles. cm^{-3}.

A relatively simple technique involves charging particles by ion attachment and subsequent trapping of particles in a filter within a Faraday cup, which is connected to a sensitive electrometer. Although this method provides a signal which needs to be calibrated, it does give a sensitive proxy of aerosol surface area, under conditions of substantial aerosol load. This is similar to the epiphaniometer which relies on attachment and detection of a radioactive lead isotope rather than an ion, and is thus sensitive to rather low concentrations of particles.

While these instruments are not themselves size selective, they can be coupled to size selecting instruments, such as the commercially available differential mobility analyser (DMA), which specifically covers the low nanometer size range. It discriminates charged particles with respect to their drift velocity under the action of an electric field. The combination of a DMA and CPC is often referred to as a scanning mobility particle sizer (SMPS). Low pressure impactors, where particles are separated by inertial impaction, can easily separate and count nanoparticles from larger particles, but commercially available systems do not provide a size resolution down to the nanometre range. An important advantage of the impactor systems is that aerosols with nanometre sizes can be collected for further analysis.

The recent developments of aerosol mass spectrometry, in which particles are vaporized and the resulting ions analysed in a mass spectrometer have provided new alternative procedures. Depending on sampling inlet configuration, size separation method, vaporization method and type of mass spectrometer coupled to it, very specific, size resolved chemical composition of nanoparticles in gas suspension can be obtained.

Direct detection of nanoparticles in liquid media faces similar physical obstacles as in the gas phase. The most successful approach into the nanometre range involves measuring the size dependent Brownian motion of an ensemble of particles through the change of interference patterns with time. Commercially available instruments reach a lower detection limit of 3 nm. Other important techniques include optical chromophore counting, resonant light scattering and Raman scattering techniques, as well as the microscopic analysis of precipitates and cross section cuts. Highly sensitive techniques within electrochemistry and mass spectrometry and Rutherford backscattering have been used to identify compounds which have been brought into tissue in the form of particles.

Most of these techniques are off-line and involve complicated sampling/sample preparation techniques. Using fluorescent molecules, quantum dots or magnetic nanoparticles as tracers, it is possible to count low concentrations of particles online. However, the choice of particles and chromophores puts further restrictions on the system studied.

Scanning Electron Microscopy (SEM) is the method of choice to investigate particle size shape and structure. When equipped with an Electron Dispersive Spectrometer (EDS) chemical composition can be determined, at least for larger particles and refractory components. The X-ray microanalysis system is not always suitable for chemical analysis because identification of substances can only be performed on the elemental level and cannot be quantified. Only solid, very high vapour pressure particles can be analysed due to the high vacuum of the system required for X-ray microanalysis.

The resolution of SEM and the related techniques has progressed below 10nm due to the implementation of cold electron sources in recent instruments. SEM resolution has been improved in Scanning Transmission Electron Microscopy (STEM) or High Resolution Transmission Electron Microscopy (HRTEM) techniques, which can again be combined to very powerful analytical techniques using electron probes and x-ray analysis.

EXPOSURE SCENARIOS

Most human individuals are routinely exposed to particles in the ambient atmosphere, primarily from diesel fumes. Any combustion process produces nanoparticles in vast numbers from condensation of gases. Initially only about 10 nm in diameter, these rapidly coalesce to produce somewhat larger aggregates of up to about 100 nm, which may remain in the air for days or weeks. The air in a normal room can contain 10,000 to 20,000 nanoparticles.cm^{-3}, whilst these figures can reach 50,000 nanoparticles.cm^{-3} in a wood and 100,000 nanoparticles.cm^{-3} in urban streets. Although the mass concentration of nanoparticles is low, it still amounts to substantial numbers. These concentrations imply that every hour, individuals breath millions of nanoparticles, and it is estimated that at least half of these reach the alveoli. At present it is not known to what extent the engineered nanoparticles contribute to these numbers. The current best practice for measuring the exposure of an individual to a material present as an aerosol is to use a personal sampling device. Samples collected are subsequently assessed either gravimetrically or via chemical analysis to determine the mass and provide an estimate of time weighted mass concentration.

Currently pollution standards are mass based; the dose by number or area will increase as size decreases. However for nanometre size aerosols, measurement of mass is not sufficient. Oberdörster *et al* showed the tremendous differences in number concentrations and surface areas for particles. The extraordinarily high number concentrations of nanoparticles per given mass may be of toxicological significance when these particles interact with cells and subcellular components. Likewise, their increased surface area per unit mass can be toxicologically important if other characteristics, such as surface chemistry and bulk chemistry, are the same.

SAMPLING

Sampling of nanoparticles is a challenging task for several reasons. First, the sampling strategy should ensure that the particle collection methods, including location, represent as accurately as possible the real exposure at the site in question and methods should be developed and chosen according to the size and nature of the particles under investigation. Secondly, because of their small mass, separation of nanoparticles from larger particles by inertial impaction can only be achieved at a relatively high pressure drop. Thirdly, considering that typical ambient atmosphere nanoparticle concentrations are less than 1 $\mu g.m^{-3}$, collection of filter samples for gravimetric analysis and chemical characterization is only feasible with certain high volume sampling techniques. In addition, the discrimination between existing ambient particles and engineered nanoparticles is an important factor in the sampling strategy.

Analysis of relatively large particles in a scanning electron microscope requires relatively little preparation of the samples. At the opposite end of the spectrum, nanometre-diameter particles to be analysed in the transmission electron microscope or scanning transmission electron microscope must be presented without contaminants on a suitably thin electron-transparent support. Inertial collection methods such as gravitational settling and centrifugal collection, are suitable only for particles greater than 1-10 μm in diameter, and are impractical to implement for nanoparticles.

Inertial deposition in impactors is achieved by increasing particle momentum in a high velocity air flow, enabling deposition onto a substrate by rapidly changing the flow direction. Use of low pressure stages in cascade impactors allows the collection of particles as small as 50 nm in devices such as the electrical low pressure impactor. Recent developments in nozzle design have led to hypersonic impactors capable of collecting particles down to 50 nm, and focusing impactors capable in principle of operating below 10 nm. However, deposition forces are necessarily high, leading to the possibility of particle damage. Aerosol samples collected by impaction are generally restricted to a small region of the substrate, thus increasing the probability of particle coincidence, and may be non-uniform with respect to particle size.

Electrostatic deposition allows relatively high deposition velocities, especially at high particle charge-to-mass ratios. Where particles are unlikely to be damaged by the charging mechanism used or the electric fields encountered, relatively gentle and uniform deposition is possible. If particles are charged to their theoretical limit, electrostatic deposition velocities are relatively independent of particle size. However, this limit is difficult to achieve under practical sampling conditions. Under conditions where positive and negative ions may freely attach to aerosol particles, a charge equilibrium is reached that is highly size dependent. The fraction of nanoparticles having a minimum of one charge drops off rapidly with decreasing size, leading to a dramatic fall in deposition velocity. Diffusional or photoelectric charging can be used to increase the average particle charge at small diameters, and electrostatic precipitation can be used effectively for particles larger than 20 nm in diameter. Below 10-20 nm, diffusion begins to dominate other deposition mechanisms. For particles smaller than 10 nm diffusion is ideally suited to obtaining uniform particle deposits on a range of sampler substrates, although samples may be highly biased towards smaller particles, and are unlikely to contain a significant fraction of particles larger than 20-30 nm.

Thermophoresis, the movement of aerosol particles in the presence of a temperature gradient, has the advantage that for a given particle composition, deposition velocity is constant below a size of around 100 nm. Achievable deposition velocities are relatively low, but deposition is gentle and unlikely to influence the physical nature of the particles. Implementation of thermophoresis in a uniform temperature gradient between two horizontal surfaces has enabled uniform deposits of discrete particles from below 5 nm to nearly 1 μm directly on to transmission electron microscope support grids.

CHEMICAL COMPOSITION OF ATMOSPHERIC NANOPARTICLES

The formation and growth of atmospheric nanoparticles, which are defined here as particles with diameters smaller than 50 nm, have been the object of intense study in recent years, due in large part to the potential role that new particle formation (NPF) may play in climate through the production and modification of cloud condensation nuclei (CCN). While models are beginning to provide information on the role that NPF plays in climate, the accuracy of such predictions is limited by our inadequate understanding of chemical processes that contribute to growth. Understanding growth mechanisms is essential because the growth rate determines whether a newly formed particle will ultimately grow to CN size (<"100 nm in diameter) or be scavenged by preexisting aerosol.

Until recently, sulfuric acid (H_2SO_4) was the only species known to contribute to nanoparticle growth. However parallel measurements of nanoparticle growth rates and H_2SO_4 show that H_2SO_4 typically accounts for only 5% to 50% of the observed growth. This suggests that other species are contributing to postnucleation growth. These compounds must possess very low volatilities in order to overcome the Kelvin effect; very few known oxidation products of common volatile organic compounds satisfy this requirement.

The first direct observation of the organic species that contribute to growth was made in the Finnish boreal forest where Mäkelä and colleagues found that nucleated particles were enriched with dimethylammonium.

Recently, we reported direct measurements of the molecular composition of 8–30 nm diameter particles formed from nucleation during the Megacity Initiative: Local and Global Research Observations (MILAGRO) field study in Tecamac, Mexico, in which we observed that

- Organic compounds contributed to 90% of the observed growth of freshly nucleated particles and
- Some of the species that contributed to this organic fraction could belong to the class of aminium compounds, cations with the structure R_3NH^+ formed by protonation of an amine, with R being an alkyl group or H.

In a subsequent article we hypothesized that amines can form organic salts with organic and inorganic acids in newly formed particles, and thus by transforming these species into ion pairs they would become essentially nonvolatile.

One conclusion from that study was that aminium salt formation may exceed ammonium salt formation when total concentrations of gas-phase amines and ammonia levels are comparable. The reactive uptake of amines into bulk aerosol has been the subject of several studies recently. One laboratory study showed that trimethylamine at <"1 ppb displaces the ammonium in ammonium nitrate particles after exposure of a few hours. An extensive laboratory study of aliphatic amines focused on the roles of both salt formation and oxidation chemistry in gas-particle partitioning. That study confirmed that alkylammonium sulfate and nitrate salts could form and that gas-phase oxidation of the amine alkyl groups can lead to low-volatility oxidation products that can physically partition into particles. Recent field measurements have shown that aminium salt formation occurs in aged organic carbon particles in Riverside, California, and in the Central Valley region of California; both of these sites are downwind of bovine sources.

Organic salt formation from the reactive uptake of amines increases the effective van't Hoff factor of the solute, thereby decreasing the water vapour saturation required for a particle to develop into a CCN. A laboratory study of the reactive uptake of ammonia onto slightly soluble organic acid particles found that this process can significantly increase the CCN activity and hygroscopic growth of these particles. Since aminium ions form similar salts with deprotonated acids this result can be extended to the reactive uptake of amines. Thus, aminium salt formation may contribute to both haze and cloud droplet formation and to nanoparticle growth. The presence of particulate organic salts was raised as a possible explanation for the relatively high aerosol hygroscopicity and cloud condensation nuclei concentrations observed in the Amazon basin.

In summary, while it is recognized that amines can form extremely low volatility compounds within particles through the formation of salts with organic and inorganic acids, the Mäkelä *et al*. study from the Finnish boreal forest and our study from MILAGRO that identified a possible aminium fragment ion have provided the only observations to date that implicate amines in postnucleation growth. This scarcity of data is due in large part to the great difficulties inherent in measuring the composition of newly formed particles, which are characterized by mass loadings on the order of a million times lower than the accumulation mode and rapid timescales for formation and loss that require analysis times on the order of minutes. This manuscript reports on recent measurements made by the Thermal Desorption Chemical Ionization Mass Spectrometer (TDCIMS), an instrument specifically designed to determine the molecular species responsible for nanoparticle growth.

We report TDCIMS and Ultrafine Hygroscopicity Tandem Differential Mobility Analyser (UHTDMA) measurements during NPF events in two very different airsheds, Mexico City and the boreal forest at Hyytiälä Finland, that directly identify aminium ions as major constituents of ambient nanoparticles. We also summarize TDCIMS observations on the abundance of aminium ions in nanoparticles formed from nucleation in Atlanta and in and around Boulder, Colorado. These observations lead us to conclude that aminium salt formation is both widespread and could account for a significant part of nanoparticle growth by organics.

We carried out laboratory measurements to characterize the response of the UHTDMA and TDCIMS instruments to 8–20 nm diameter aminium salt particles formed from the following model compounds: methylamine (MA), dimethylamine (DMA), trimethylamine (TMA), acetic acid (AA), and propanoic acid (PA).

Because theory predicts that for a given material and in the absence of significant curvature effects, the diameter decrease due to heating should be independent of size for particles that are much smaller than the mean free path of air, we express volatility as the decrease in particle diameter after heat treatment. The hygroscopicities of the model salt particles all fall within a narrow range of values, which show a dependence on diameter due to the Kelvin effect. In general the amine salts are slightly less hygroscopic than ammonium sulfate, *e.g.*, 1.20 ± 0.05 at 10 nm compared to 1.35 for ammonium sulfate at the same diameter.

The same model salt particles were analysed with the TDCIMS to assess its ability to detect and quantify particulate amines, organic acids, and aminium salts. Regarding the organic acids, a recent study of the TDCIMS response to laboratory-generated four- to eight-carbon monocarboxylic and dicarboxylic acid particles concluded that the instrument could identify each by their deprotonated parent ion with minimal fragmentation, at sensitivities of *ca.*100 Hz of integrated peak area per pg of collected aerosol mass. This is approximately equal to the instrument sensitivity towards sulfate in laboratory-generated ammonium sulfate aerosol.

The main results obtained from these laboratory experiments were as follows:

- Individually, particulate amines and organic acids are both observed in the positive ion spectra as protonated parent ions, while in negative ion spectra organic acids are observed as deprotonated parent ions;
- Aminium salt particles are desorbed and ionized within the tdcims as if the amine and acid were independent from each other, that is, additional reactions between the desorbed species within the ion source such as those resulting in amide formation are not observed;
- Tdcims sensitivity towards particulate amines is equivalent to that of particulate organic acids. These studies, and the observations reported below, are based on reported ion abundances.

While our results can be extended to particulate volume and thus actual diameter growth rates by estimating the average molar volumes of the aminium salts, this would require additional laboratory evaluations that are beyond the scope of our current investigations.

For this study we can state more generally that a significant ion fraction of aminium salts implies a significant volume fraction, and thus a significant impact on nanoparticle growth.

We begin with a detailed analysis of two NPF events from two contrasting locales. The first event took place in Tecamac, Mexico (19.703N latitude, 98.982W longitude, 2273 m altitude), a mixed rural/urban site situated 40 km N of central Mexico City and heavily influenced by local rural sources as well as regional emissions, on March 21, during the MILAGRO field campaign. The diameter growth rate for this event was 20 nm hr^{-1}.

This growth rate was used along with measurements of gaseous H_2SO_4 to estimate that H_2SO_4 contributed about 8% of the observed growth, a result that is consistent with nanoparticle composition measurements for a similar event that took place at the same location 4 d earlier. Therefore we can conclude from this analysis that organic compounds are responsible for over 90% of the growth observed in the Tecamac event.

FATE AND BEHAVIOUR OF ATMOSPHERIC NANOPARTICLES

MEASUREMENTS AND APPROACH

Discrepancies in the HO_x budget will be explored by expanding the suite of measured species at the PROPHET site, integrating these measurements into box models and comparing the results to the studies described above.

This comprehensive approach will require measurements of OH, $HO_2$and OH reactivity in addition to measurements of actinic flux, NO_x, CO, O_3, HONO, VOC (including aldehydes) and VOC oxidation products (OVOC), such as formaldehyde (HCHO), methacrolein (MACR), methyl vinyl ketone (MVK), organic nitrates and PANs, and glyoxal. Measurements of OH and HO_2 will be made by LIF using the Fluorescence Assay by Gas Expansion (FAGE) technique both at the top of the PROPHET tower and near the forest floor, and will be collocated with measurements that are required for the box model.

A full range of volatile terpene and other VOC concentrations and fluxes will be measured, which will complement the existing record of isoprene flux data collected from 1997-2007. BVOC measurements will be made at the PROPHET site from aircraft and at the canopy, branch, and soil levels. Previous studies of emissions from the dominant vegetation at UMBS identified isoprene, MT and SQT as key emitted species.

Light- and temperature-dependent MT emissions were significant and approximately one order of magnitude greater than solely temperature-dependent MT emissions, but a box model analysis confirmed that isoprene dominates OH reactivity, especially during daytime hours. Canopy-scale fluxes of isoprene, total MT, and selected oxidized VOC (OVOC) will be measured using a high sensitivity PTR-MS and disjunct eddy covariance (DEC). Canopy scale fluxes of isoprene, MT, and SQT will also be made by relaxed eddy accumulation (REA). In addition to these ground-based measurements, fluxes of BVOC will be made by collecting samples aboard a small aircraft laboratory (ALAR) and by disjunct eddy accumulation (DEA). Continuous measurement of O_3 at three tower heights (2m, 10m and 33m) via independent sampling lines and dedicated instruments will provide continuous O_3 vertical profiles.

Ambient concentrations of speciated VOC will be measured via in-situ sampling and analysis by GC-MS. In addition, the measurement mode of the PTR-MS will be alternated between performing canopy scale flux measurements and performing mass scans at different heights on the tower as a means of comparison to the GC-MS. The mass scans will determine mixing ratios of a wider range of species including primary and secondary photoproducts of isoprene and MT. Branch-level emissions through the canopy will be measured using vegetation enclosure systems with chemical analysis techniques relying on solid adsorbent sampling and GC-MS analysis. These enclosure studies will allow comparison between emissions of reactive BVOC with canopy scale flux measurements. Reactive BVOC such as SQT will be oxidized rapidly within the canopy and are not readily detected from tower-based measurements where samples have to be brought to instruments through sampling tubing. However, it is possible that gradient and flux measurements of OVOC (in particular HCHO) could support the implied very fast oxidation processes.

These measurements will be analysed to address the following sub-questions regarding the impact of BVOC emissions on HO_x chemical cycling, sources, and sinks:

How Important is the Contribution of Enhanced Cycling of Radicals to the Observed Discrepancies in the HO_x Budget

Recent measurements of HO_x radicals using LIF over the Amazon forest were found to be greater than expected, and it was suggested that BVOC oxidation kinetics under low NO_x conditions may efficiently recycle HO_x radicals leading to higher than expected concentrations of OH. This recycling may involve production of OH from reactions previously thought to terminate HO_x radicals, such as the reactions of HO_2 with RO_2 radicals. Conditions at PROPHET are similar and these new radical cycling reactions could lead to concentrations of OH radicals higher than currently modeled. This could imply that forest emissions are not as significant a sink of OH radicals on the global scale as previously believed.

The measurements described above will be used in both 0-D and 1-D models to predict HO_xconcentrations during the day and night. An analysis of these results will help to determine whether conditions for HO_x photochemistry at PROPHET have changed since 1998 and whether a more complete characterization of BVOC has improved the model/ measurement agreement. Analysis of updates to the chemical mechanisms of BVOC oxidation, including the chemistry of higher generation oxidation products, in addition to postulated changes in the efficiency of radical propagation by $HO_2 + RO_2$ reactions will help to determine whether an incomplete description of radical chemistry in forest environments is responsible for any measurement/model disagreements.

What is the HO_x Production Rate from Photolysis of Aldehydes and HONO

Measurements of HCHO, acetaldehyde, HONO and actinic flux will be made to estimate the contribution to HO_x production from these species in addition to O_3 photolysis. There are significant emissions of acetaldehyde from vegetation at PROPHET. Primary emissions of HCHO are not known but it is produced as a primary photoproduct from OH-initiated oxidation of isoprene.

There is recent evidence that HONO is produced both during the day and night from surface chemistry and its photolysis has been found to be a significant source of OH radicals in low-NO_x rural environments. Within the forest canopy, HONO may become a major photolytic precursor over O_3 and HCHO, due to its broad absorption in the UV-A light

and the enhancement of its heterogeneous production on available surfaces. Measurements of HCHO will be made by LIF, measurements of HONO will be made by aqueous scrubbing derivatization, and measurements of acetaldehyde by PTR-MS.

Accurate determinations of photolysis frequencies above- and within-canopy are crucial to understanding fast HO_x photochemistry. Vertical light distribution through a forest has been estimated by radiometer measurements at two levels in Germany. However, the uneven structure of the UMBS forest in general and the unknown future structure of the FASET forest suggest that a more extensive horizontal investigation of light penetration be conducted. In addition to spectral radiometers above the forest canopy, a network of portable UV sensors on the top of and around the base of both the PROPHET and FASET towers will be deployed. In addition, transects up to 1 km away from each tower will be periodically walked with handheld instruments to capture large-scale heterogeneity within the tower footprints.

Could the Ozonolysis of Unknown Reactive Terpenes be a Significant Source of HO_x Radicals

Another possible reason for some of the observed HO_x discrepancies described above may be the ozonolysis of heretofore unmeasured reactive BVOCs that act as a net OH source via alkene reactions with O_3. These highly reactive species are suspected to play significant roles in modifying O_3 and OH concentrations within the canopy. Previous measurements of O_3 uptake within the canopy at the Blodgett forest site suggest that gas-phase chemistry dominates the loss of O_3, implying that there exists an unrecognized source of reactive compounds. The ozonolysis of these emissions could lead to significant OH radical concentrations inside the forest canopy.

To investigate this, the total BVOC reactivity with O_3 ($\bullet k_{O3i} \bullet [VOC]_i$) will be determined by conducting vegetation enclosure experiments. Instead of eliminating $O_3$from the purge

air, air flowing into the enclosure will be adjusted to ambient O_3 levels using a series of O_3 monitors and an in-line O_3 generator. This will allow a direct measurement of the difference in O_3 levels between the purge air flow and the air flow exiting the enclosure via the sampling flow. Modified TEI Model 49C O_3 monitors have yielded precision in 1-min O_3 gradients < 0.10 ppb for direct O_3 gradient measurements. Given O_3 lifetimes with respect to reactive BVOCs on the order of 45 minutes at ambient O_3 levels and residence times of 3-5 min inside the enclosures, O_3 reaction losses inside the enclosure of several ppb are expected. The sensitivity of this proposed O_3 differential measurement will be sufficient to investigate the overall O_3 reactivity of the emissions. Control experiments will be conducted using empty enclosures to determine the O_3 behaviour from the experimental materials alone. Vegetation enclosure experiments will be conducted over several diurnal cycles and the day/night and temperature-dependent changes of the measured O_3 reactivity will be determined.

Temperature behaviour and nighttime O_3 losses will be used to assess the O_3 loss to surfaces (*i.e.* leaves, stems) of enclosed vegetation and ground cover. BVOC emissions identified in standard enclosure experiment under removal of O_3 and estimated O_3 reaction rate constants will be used to determine the O_3 reactivity that is accounted for, and these data will then be compared with the directly measured total O_3 reactivity. The agreement/discrepancy between these determinations will then be used to assess the fraction of O_3 reactivity that cannot be accounted for by currently identified BVOC emissions. Nighttime measurements of OH and HO_2 will help assess the "dark" source of oxidants to compare with measured rates from O_3 + terpene reactions near the surface and at tower height.

Can the Missing OH Reactivity be Attributed to Specific Plant Emissions

Measurements of total OH reactivity using a turbulent flow tube technique coupled with LIF detection of OH will be made and compared to calculated reactivity from measured gas phase species concentrations (notably: CO, NO_2, HCHO, isoprene, MT, + other VOC). Recently the NCAR Biosphere-Atmosphere Interactions group has identified a BVOC compound, methyl salicylate, with emission characteristics thought to be similar to MT. Ambient concentrations and emissions rates of this compound will be determined by the NCAR group during PROPHET 2009 to help assess whether this compound could account for some of the missing BVOC reactivity.

To further investigate the nature of missing VOC reactivity a new type of experiment will be conducted to measure the OH reactivity of BVOC emissions holistically at the branch level for various plant species. This experiment will determine missing OH reactivity by comparing calculated OH loss frequencies from measured speciated VOCs in the enclosure to measured OH and O_3 loss frequencies. OH reactivity of compounds emitted from vegetation, soils and leaf litter will be measured using the approach described by Sinha *et al.* who report an OH reactivity lower detection limit of ~6 s^{-1}. This approach has been improved.

A reactive VOC tracer is passed through a glass reactor and its concentration is measured with a fast-response PTR-MS. OH radicals are introduced in the glass reactor at a constant rate to react with the VOC, first in the presence of zero air and then in the presence of ambient air. Comparing the amount of the VOC tracer exiting the reactor with and without the ambient air allows the OH reactivity of the air to be determined. These measurements can then be compared to ambient measurements of OH reactivity using flow tube techniques with LIF detection of OH described above.

ATMOSPHERIC CONCENTRATIONS

CARBON DIOXIDE

The fundamental reason why scientists are generally confident that we are and soon will be experiencing global warming is because the concentration of carbon dioxide (CO_2)

has been increasing steadily in the global atmosphere (Fig 5.7). Unlike water vapour or aerosol, CO_2 is well-mixed in the atmosphere and changes are relatively slow. There is an annual cycle, evident in Fig 5.7 and explained, but its amplitude is small compared to the trend.

(The annual cycle of water vapour concentration is much larger than any trend.) Carbon dioxide is a major greenhouse gas, and even the simplest one-dimensional radiative transfer model predicts surface warming when greenhouse-gas concentrations rise.

Other factors affecting climate, such as the direct and indirect effect of aerosols, are much harder to measure, display large variability, and their effect on climate is poorly understood. It is because of these 'other factors' that scientists are not overly confident. Estimates of the atmospheric CO_2 concentrations prior to 1958 can be made by analysing the bubbles in ice at different depths in ice sheets, for instance in Antarctica. This analysis shows that there were about 280 parts per million of CO_2 in the Earth's atmosphere prior to 1800 AD, at least as far back as 1000 AD.

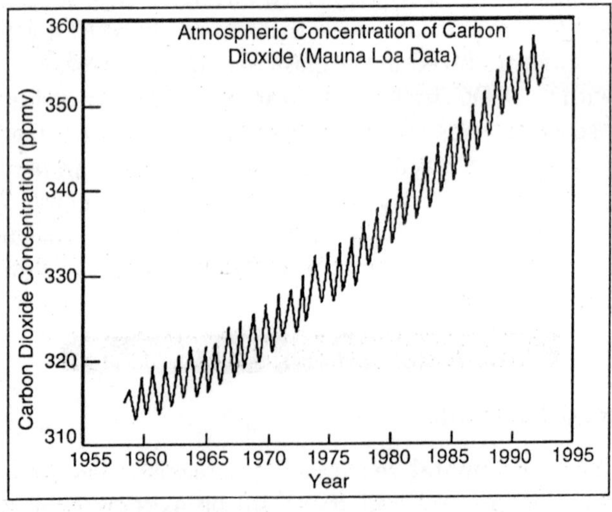

Fig. 5.7 Atmospheric CO_2 Concentrations.

So the concentration was fairly steady until about 200 years ago, and it started to rise during the industrial revolution. And the surplus CO_2accumulated during the last 40 years is almost twice the surplus built up during the previous 150 years. So during the last two centuries the growth rate has increased. It can be seen from Fig 5.8 that the recent rate of increase has been fairly steady, but closer inspection reveals slight variations.

The rate of increase itself is shown in Fig. 5.8 The rate was about 0.7 ppm annually in the early 1960's, but it had more than doubled by 1994. There was a decline from 1.7 ppm/a around 1987 to 1.3 ppm/a five years later, but a renewed increase since then.

More recent evidence suggests that that this renewed acceleration is short-lived, and that the growth rate now is steady around 1.6 ppm/a. Many developed countries, especially in Europe, have stepped up measures to reduce fossil-fuel consumption and any other industrial/agricultural processes that contribute to greenhouse gas emissions. But overproduction of fossil fuels, especially oil, has thwarted conservation efforts.

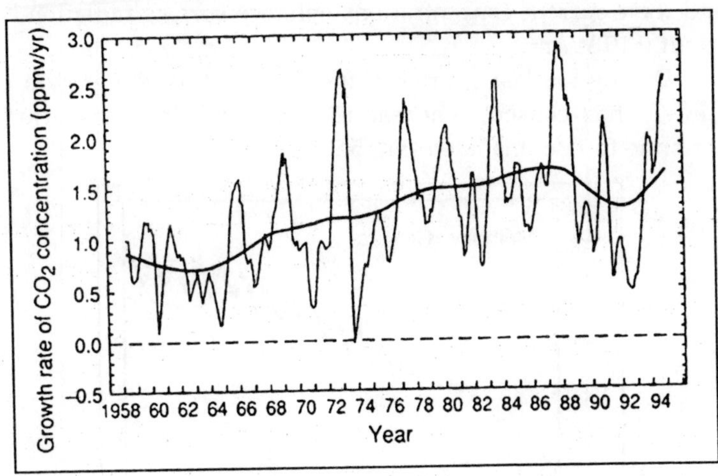

Fig. 5.8 Annual Growth Rate of Atmospheric CO_2 Concentrations at Mauna Loa.

The rate of CO_2 increase appears to be slowed after El Nio events, such as the 1982-83 and the 1997 events. This is presumably because the global mean temperature is slightly higher during El Nio years, and therefore plants grow more prolifically, especially at boreal latitudes. This implies more CO_2 uptake by the biosphere.

It is not clear, however, why there is a two-year gap between the El Nio event and the reduction of atmospheric CO_2 growth rate. A key factor is the rate of uptake by the ocean.

GREENHOUSE GASES

The concentration of CO_2 has increased by about 30% since the beginning of the industrial revolution. By comparison, two other atmospheric gases of major importance in greenhouse warming (apart from water vapour, which is dominant), methane (NH_4) and nitrous oxide (N_2O) have increased by about 145% and 15% respectively.

The present concentrations lead to enhanced radiation to the ground by 1.56 W/m^2 from the CO_2, 0.47 from NH_4 and 0.14 from the N_2O. Chlorofluorocarbons did not exist before 1950, and their current concentrations enhance surface radiation by about 0.10 W/m^2.

During the last quarter of the 20th century the growth of CFC-11 has ceased, whereas nitrous oxide concentrations continue to rise unabated (Fig 5.9).

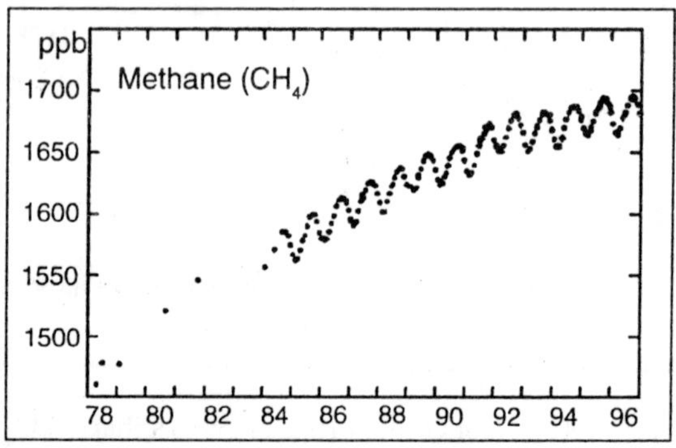

Fig. 5.9

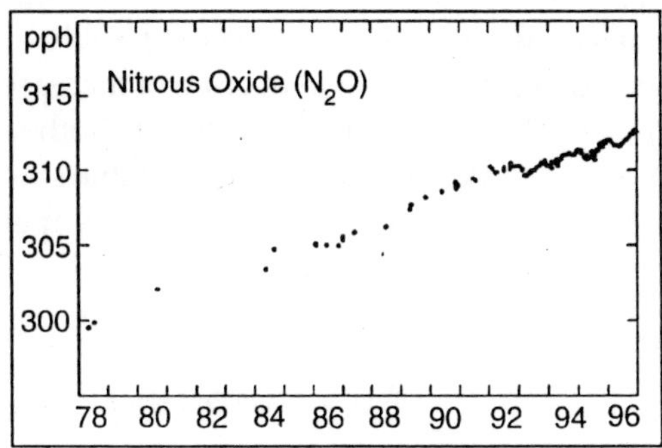

Fig. 5.10

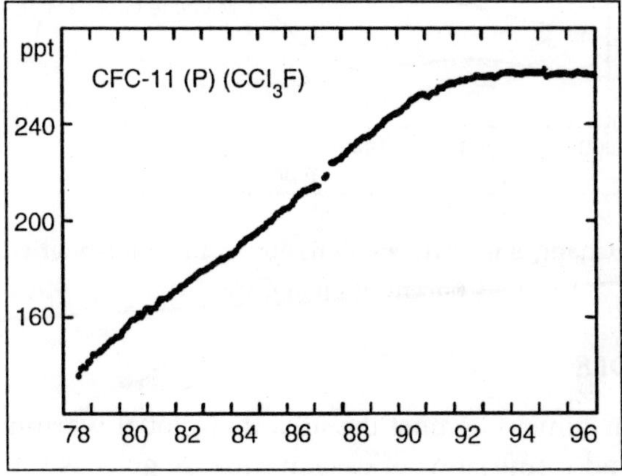

Fig. 5.11 Changes in Concentration of Methane, Nitrous Oxide and CFC-11.

The rate of increase of methane has been largest, which is especially important since a NH_4 molecule has 25 times the capacity of a CO_2 molecule to trap heat in the atmosphere. The pre-industrial methane concentration was 0.75 ppm and the 1997 level was 1.73 parts per million. The main culprit for its growth appears to be agricultural activities, in particular rice cultivation (anaerobic decomposition in paddies) and boviculture (enteric fermentation). More recently, the NH_4 growth rate appears to have declined, particularly in the northern hemisphere (6). This is because emissions have decreased considerably.

There was also a slowdown in the rate of increase of nitrous oxide (N_2O) in 1992, compared with 1985-91. Both natural and plowed soils are an important but highly variable source of CO_2, CH_4, and N_2O, depending on soil moisture, temperature and fertilizers.

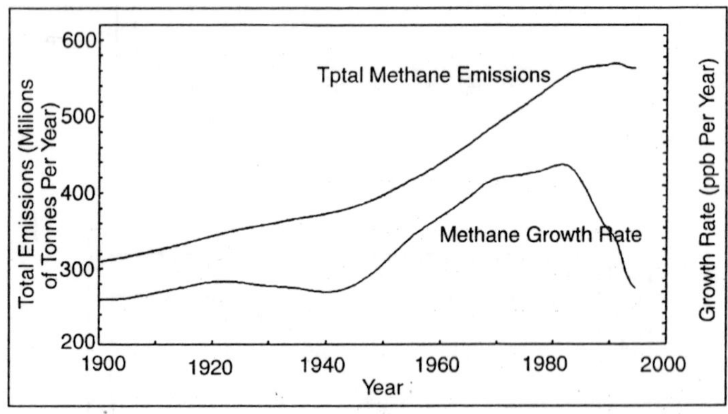

Fig. 5.12 Changes in Methane Emissions and Atmospheric Methane Concentration at Cape Grim

AEROSOLS

An argument against the idea that global warming is due to mankind's emissions of carbon dioxide goes as follows: the warming this century occurred mostly *between 1910 and 1940,* when the carbon dioxide concentration grew slowly from 293 to 300 ppm. On the other hand, the temperature remained steady between 1940-1980, while the carbon dioxide concentration increased from 300 to 335 ppm.

The most likely answer to this inconsistency is atmospheric aerosol. Aerosols are emitted by industrial processes, transport, etc, and their increased concentration offset simultaneous warming due to increasing greenhouse gases. However the warming overtook the cooling by mid 1970's. Tropospheric aerosols reduced solar radiation to the ground by about

0.5 W/m^2, as a global average, between 1940-1992. Unlike greenhouse gases, which are generally long-lived, aerosols fall out of the atmosphere fairly rapidly, either dry ('sedimentation') or within rain (as condensation nuclei).

Therefore aerosol concentrations are not uniformly mixed across the globe. They have been so high in some regions that the cooling effect may have exceeded the warming due to greenhouse gases. In fact the lack of warming between 1940-1980 is only found in the northern hemisphere, where most manmade aerosols are emitted.

6

Analysis and Characterization of Nanoparticles in Aquatic Environments

INTRODUCTION

The increasing use of engineered nanoparticles in research and product development in application areas related to medicine, sensing, environmental and consumer products engenders a growing need to understand their properties and behaviours as they are synthesized, applied, and evolve (or age) in a particular environment, process, or application.

There is also an increasing awareness of the need to understand the health, safety, and environmental impacts of nanoparticles in both their synthesized form and as they evolve through application or environmental interaction.

Although novel and unusual properties of nanoparticles and other nanostructured materials excite scientists, technologists, and, often, the general public, the sometimes surprising properties of many of these materials raise analysis and characterization issues that can also be unexpected by analysts, scientists, and production engineers.

It is increasingly recognized that reports on the properties of nanoparticles and other nanostructured materials are sometimes based on inadequate characterization. Consequently, the validity of some of the conclusions may be questionable. The major objectives of this paper are to identify the subset of important information that can be obtained about nanoparticles using tools under the general heading of surface chemical analysis and to examine some of the issues and challenges faced when performing such analyses.

It is widely recognized that as particle size decreases to the nanometer scale, there are a variety of reasons, including quantum confinement effects, that cause their physical and chemical properties to differ from those associated with their bulk form. Equally important and widely acknowledged, but seemingly less understood, is recognition that a large portion of the atoms in nanoparticles are at or near the surface of the particles.

Grainger and Castner point out that over the past 40 or more years, surface scientists have obtained detailed knowledge about the behaviour of surfaces, including the important role of deliberate and accidental surface layers, which has lead to the development of a set of tools that can be used to understand and characterize surfaces.

They further argue that the same rigor that has been applied to surface studies is needed (but, with few exceptions, is usually not used) to understand and control the properties of nanoparticles.

They called this nanosurface analysis. For example, nanosurface analysis is extensively used to characterize supported nanoparticle catalysts, but virtually unused to characterize unsupported nanoparticles in biomedical applications.

In a different report, Karakoti *et al.* noted that the importance of nanoparticle surface chemistry, especially as applied to toxicity, has been (surprisingly) underemphasized.

A March 2006 article in *Small Times* magazine described a workshop designed to identify roadblocks to nanobiotechnology commercialization In this article, several experts opined that many of the important physical characteristics needed to understand the physical and chemical properties of nanoparticles are unreported in research reports and apparently often unmeasured. This was found to be especially true in areas related to assessments of particle toxicity.

The article further notes that the changes these particles undergo when placed in various environments for storage or use are especially important and usually unknown. In many cases, nanoparticles are coated with surfactants or contaminants, and these are often not well characterized and, many times, not even adequately identified.

Based on these and other articles, it appears two aspects of nanoparticle properties are not always fully appreciated. These include the importance that the nanoparticle surface composition and structure play on their properties and performance and how significantly these properties can change with time and environmental exposure.

Because of the increasing importance of nanoparticle characterization, working groups of the International Bureau of Weights of Measures (BIPM), Consultative Committee for Amount of Substance: metrology in chemistry (CCQM), and International Organization for Standards (ISO) Technical Committee (TC) 229 on Nanotechnology are focusing considerable attention on nanomaterials characterization. Specific working groups are focused on the characterization of nanoparticles for Environmental Health and Safety (EHS) issues and for toxicology studies.

Surface characterization is a subset of several analysis needs, and the surface characterization needs of ISO TC 229 are being addressed in a technical report (TR14187) being prepared by the ISO TC 201 Committee on Surface Chemical Analysis. This article introduces some of the topics and issues expected to be included in ISO TR14187.

A Working Party on Manufactured Nanomaterials (WPMN) of the Organization for Economic Co-operation and Development (OECD) has established a list of physical-chemical properties and material characterization needs for nanostructured materials. Among the 16 properties on this list are characteristics related to surface chemistry and surface charge. Although there are many different types of information needs for nanoparticles and other nanostructured materials, ISO TC229 Joint Working Group (JWG) 3 has tentatively focused on approximately seven of the measurement criteria that emphasize the importance of surface chemistry.

These criteria include properties of surface layers and surface contamination, the chemical state or enrichment of species on the surface or at interfaces, and information about surface functionality.

These examples relate to nanoparticle synthesis ($LuPO_4$ in apoferritin), environmental-induced changes in the nanoparticles (iron nanoparticles), and confirmation of the desired property changes in chemically functionalized nanoparticles (functionalized carbon nanotubes [CNTs]). The $LuPO_4$ particles formed inside an apoferritin template have potential application in cancer radioimmunotherapy and radioimmunoimaging.

One analysis need for these particles was to confirm the presence of the Lu in the phosphate form inside the aproferritin shell. This was accomplished using XPS.

Zero valent iron nanoparticles have been used to reduce environmental contaminants such as CCl_4. In nanoparticle form, they sometimes exhibit beneficial reaction pathways relative to micron-sized particles and can also be delivered by injection into contaminated wells.

We have found that iron nanoparticles with a metal-core/oxide-shell structure are more reactive than particles with iron only in +2 or +3 states. However, not all metal-core/oxide-shell iron particles produce the same chemical pathways.

It has been found that "bulk" tools, such as X-ray diffraction (XRD), and high-resolution methods, such as transmission electron microscopy (TEM), combined with surface methods, including XPS, are all important for detailed characterization of these particles.

In recent work, the impact of natural organic coatings formed on the particles in solution has been studied. XPS, as well as optical methods, again have been useful for characterizing these coatings. CNTs in many different forms are used for a variety of different technologies. For many applications, including sensor technologies, it is beneficial to functionalize their surface. In the example in Figure 5.12, a succinimidyl ester is adsorbed onto the CNT surface. Amine groups on a protein can react with the anchored succinimidyl ester to immobilize proteins on the surface.

Possible applications of this approach include the creation of biosensors and the possibility of using such interactions to guide self assembly of CNTs. For this example, XPS, in combination with high-resolution electron microscopy, provides important information about the presence and nature of the molecules attached to the surface.

As the importance of nanoparticle surfaces is recognized, it would seem logical that tools developed to extract information about surfaces could and should be applied to nanomaterials. At least two different issues appear to limit both the attempted applications of these methods and the impact of these tools in nanoparticle characterization.

First, many of the tools appear to lack the necessary degree of spatial resolution in three dimensions needed to analyse individual nanoparticles (or the variations of composition within the particles). Consequently, some researchers may not consider application of the tools even though they can often provide essential information that cannot be obtained by other means. Second, the tools are sometimes applied to nanostructured materials without considering a range of analysis challenges or issues that these materials present. This can lead to inconsistent or even wrong conclusions.

Grainger and Castner and Baer *et al.* have pointed out several important characteristics of appropriate nanoparticle analysis including: analysis during and immediately after synthesis, understanding the interactions (and time dependence) of particles when placed in the application environment (biological, solution, catalysis), and the necessity of multi-technique analysis. Although this paper focuses primarily on the application of surface analysis tools, such tools provide only an important (and often ignored) subset of the information needed to understand and control the behaviour of nanoparticles.

This paper addresses two important issues. First, the types of information that can be gained from application of surface analysis methods, and, second, some of the technical challenges faced when applying surface analysis tools (and often other tools) to nanoparticle characterization.

SURFACE CHEMICAL ANALYSIS

Although many different surface-analysis techniques can be useful for the characterization of nanoparticle surfaces, this paper will focus on information that can be obtained from electron spectroscopies (Auger electron spectroscopy [AES] and X-ray photoelectron spectroscopy [XPS]); ion-based methods (secondary ion mass spectrometry [TOF-SIMS] and low energy ion scattering [LEIS]); and scanning probe microscopy (SPM), including atomic force microscopy (AFM) and scanning tunneling microscopy (STM).

A considerable amount of effort has been devoted towards the development of guides and standards for AES, XPS, SIMS, and SPM methods since they are primary surface analytical methods. This includes work by Standards Committees ASTM E42 on Surface Analysis and ISO TC201on Surface Chemical Analysis. The second ion-based method, LEIS, has been added to this list because of its extreme surface sensitivity that can provide useful information for nanoparticle analysis. It can be argued that SPM-based methods, which allow visualization of nanostructures on surfaces, have been a significant driver for the development of nanotechnology.

Finally, literature reports suggest that XPS has evolved into the most widely used surface spectroscopy. Detailed discussions of these methods are available from many sources. The primary objective of the short descriptions that follow is to highlight features particularly useful for nanoparticle analysis. Although these are generally labeled surface analysis methods, each technique has nanometer resolution in at least one dimension.

Table. Lists Some Types of Information that can be Extracted from Nanoparticles Using Surface Analysis Tools Along with Several Characteristics of the Methods.

Electron Spectroscopies	Information available	Probe	Detected	Lateral Resolution	Information Depth	Depth Resolution
Auger electron spectroscopy (AES)	• Surface composition of individual large nanoparticles or distribution of smaller nanoparticles (depending on spatial resulution of specific instrument). • Enrichment or depletion of elements at surface • Presence and/or thickness of coatings and/or contaminants	Electrons (~ 3 to 20 kV)	Auger electrons	≈ 10 nm	≈ 10 nm	≈ 2 nm
X-ray photoelectron spectroscopy (XPS)	• Analysis of a collection of particles deposited on a substrate or other support. • Surface composition and chemical state • Presence and nature of functional groups on the surface • Enrichment or depletion of elements at surface • Presence and/or thickness of coatings or contaminats • Nanoparticle Size (when smaller than ~ 10 nm, can sometimes determine average particle size when too small to be detectred by other methods or in complex matrix.) • Electrical properties of nanopartlcles and coatings	X-rays	Photoelectrons	≈ 2 μm	≈ 10 nm	≈ 2 nm

Incident Ion Methods						
Secondary Ion Mass Spectrometery (SIMS)	• Usually analysis of a collection of particles or larger individual particles deposited on a supporting substrate. • Presence of surface coatings or contamiinants on collections of nanoparticles • Functional groups on surface	Ions (~ 3 - 20 kV)	Sputtered ions	≈ 50 nm (inorganic) > 200 nm (organic)	≈ 1 nm	≈ 1 nm (inorganic) ≈ 10 nm (organic)
Low energy ion scattering (LEIS)	• Presence of ultra thin coating or contamination • Effects of size	Ions (~ 2 to 10 kV)	Elastically scattered ions	≈ 100 μm	≈ 10 nm	≈ 0.2 nm
Scanning Probe Microscopies						
Scanning Tunneling Microscopy (STM)	• Electrical characteristics of individual nanoparticles • Nanoparticle formation and/or size distribution of particles deposited or grown on a surface	Stylus	Tunneling current	≈ 1 nm	≈ 10 nm	
Atomic force microscopy (AFM)	• Shape, texture and roughness of individual particles and their distribution for an assembly of particles • When particle structure is known, can provide information about crystalligraphic orientation	Stylus	Force or displacement	≈ 1 nm	≈ 10 nm	

It is important to recognize that the different surface analysis techniques can provide complementary and comprehensive information as highlighted in a bubble chart developed at the UK National Physical Laboratory (NPL). This chart summarizes the types of information that can be provided by each of several different analysis methods. The analysis methods in the NPL chart go beyond those discussed here, and many may be useful for nanoparticle characterization. The types of information that can be obtained include topography, elemental composition, molecular and chemical state, and structure.

The following paragraphs summarize the essence of five surface analysis methods and provide a few examples of their use. These examples offer a sample, but not a comprehensive picture, of what can be done.

Applications are limited only by the ingenuity of the research team involved. It will be apparent that application of these tools is most often useful when they are applied in combination with other tools providing complementary information (*i.e., multi-technique analysis*). We have found, for example, from studies on 50 nm iron metal-core/oxide-shell nanoparticles that XPS, TEM, and XRD are standard measurements that should be applied to particles from most experimental tests.

Because of environmental and time (or processing induced) changes in the particles, we make many measurements after exposure to different environments and after any significant time of particle storage. Whenever possible, we apply tools that work in the environment of interest (*in situ*measurements) and attempt to compare results for consistency. When materials are removed from solution for *ex situ* analysis, we usually carefully dry them and handle them without exposure to air before analysis.

ELECTRON SPECTROSCOPIES

Both AES and XPS involve the detection of electrons emitted from samples with kinetic energies typically below 2000 keV. Although a wider range of energies are available when synchrotron X-ray sources are used for XPS, most of the discussion will focus on laboratory-based sources. Much of the value of these methods is their surface sensitivity that arises from the short distances that electrons travel at these energies without undergoing inelastic scattering and energy loss.

Therefore, the electrons detected in Auger or photoelectron peaks are from the outer few nanometers of the material, as indicated in Table.

Catalysis was one of the first areas where the combination of "bulk" analysis methods with these surface analysis methods allowed information about the enrichment or depletion of elements on the surface to be determined. Because electrons that emerge from the material that have lost energy appear in the background region of the spectra, it is possible to use these methods to provide depth, enrichment, or layering information within the XPS and AES analysis volume.

Consequently, these two methods can be used with multiple approaches to obtain important information about layering or coatings on particle and nanoparticle surfaces.

Although both X-ray and electron excitations produce Auger electrons, AES is usually associated with the use of incident electrons to generate Auger electrons. These incident electrons typically range in energy from 2 to 20 keV. As might be expected, X-rays (often Mg or Al Kα) are the incident radiation in XPS. This technique is also referred to as Electron Spectroscopy for Chemical Analysis (ESCA)—coined by Kai Siegbahn, who was awarded a Nobel Prize for his development of this technique.

Because an electron beam can be focused to less than 10 nanometers in size, it is possible to analyse individual nanoparticles with AES. However, issues related to electron penetration and scattering can result in worse resolution than expected based on beam size alone.

Although XPS does not have the spatial resolution to analyse individual nanoparticles (with the possible exception of a few special synchrotron-based systems with a highly focused and bright source of X-rays), it is often possible to analyse collections of particles (in a single layer or, effectively, in powder form) and to obtain useful information. Both AES and XPS can be extremely important tools for determining the presence, composition, and thickness of coatings on nanoparticles, as well as surface enrichment and depletion at particle surfaces.

To the surprise of many, XPS can sometimes be used to determine particle sizes when conditions are not appropriate for analysis by other methods. The size, shape, and layered structure of nanoparticles influence XPS data in several different ways including:

- Peak intensities and relative peak intensities of
  - Peaks for different elements
  - Different peaks for the same element
  - Same peaks excited with different X-ray energies
- Peak energies
  - Binding energies of peaks
  - Value of the Auger parameter
- Background signals from electrons that have lost energy

Depending on what is known about the specimen and the analysis objective, each of these influences can be used to extract useful information about nanostructured samples.

Although electron spectroscopies may not be as used in every situation where they would be appropriate, they are increasingly employed (especially XPS) for nanoparticle characterization in a variety of ways. It is valuable to identify specific analysis objectives based on the type of research involved.

Understanding the analysis objectives assists both the data collection and the nature of the data processing. This can significantly improve the quality and value of the resulting information.

One simple example of how solution composition alters the surface of a nanoparticle is provided before summarizing several other examples from the literature.

The topics covered include:

- Intended or unintended coatings, as previously noted, can have a significant impact on a variety of nanoparticle properties. The influence of natural organic material (NOM) on the reactivity of iron metal-core/oxide-shell nanoparticles in deionized (DI) water serves as an example of both the formation of coatings and their impact. Johnson *et al* demonstrated that when NOM was added to water, the nanoparticles were less likely to be retained in a soil column. Previous reactivity and microscopy work has shown how the particles change properties in solution as a function of time. We have started an examination of the influence of waterborne NOM on the rate of particle change. Although the studies involve many parts, we have used XPS to examine changes in the particle surface chemistry, TEM to examine morphology changes, and XRD to examine the overall oxidation of the particles as a function of time. An overview of these measurements is provided in Figures 6.1. The TEM images show the types of changes that occur in solution over a day. The oxide shell changes from highly ordered to something more nodular.

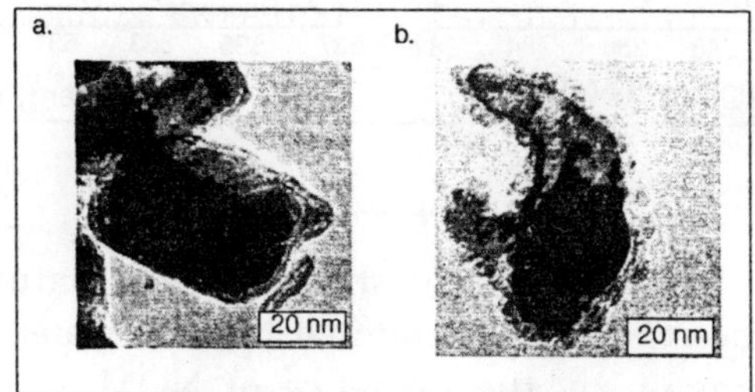

Fig. 6.1

The XPS measurements in Figure 6.1 show that the Fe near the surface starts as Fe^{+3}, but a Fe^{+2} component is formed after exposure to the solution.

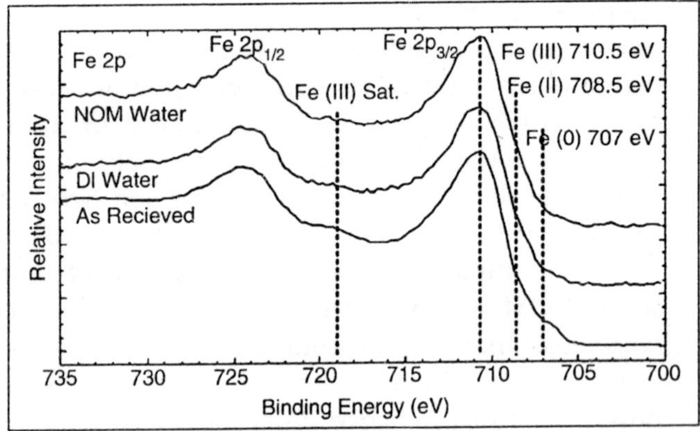

Fig. 6.2

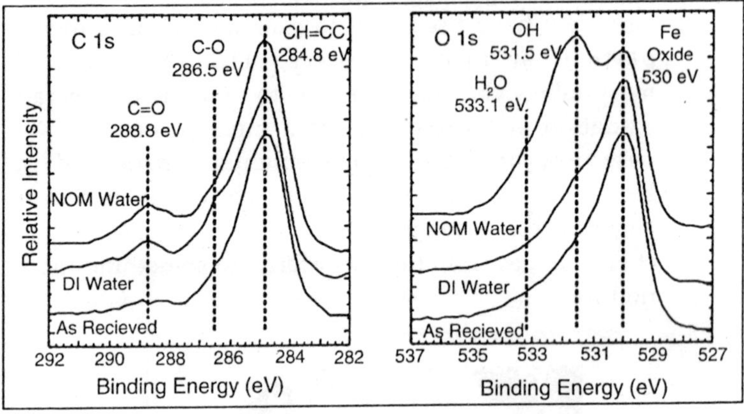

Fig. 6.3

The XPS results also show that the nature of the surface C and O are altered by both water exposure and again by the presence of NOM. Carbon is a ubiquitous feature of materials exposed to the atmosphere, but the addition of NOM thickens the carbon layer on the particles, as well as increasing the O to Fe ratio. Note that it is possible using the approach of Smith or that of Shard *et al.* to estimate the thickness of a carbon overlayer on the particles. Although there are additional complications because of O in the NOM, the Smith approach has been used to estimate carbon layer thickness in Table.

	Initial	H_2O	H_2O+NOM
Fe [at%]	30	16	10
O [at%]	42	53	47
C [at%]	28	31	43
Approximate C thickness [nm]	0.87	0.98	1.5

The presence of the NOM increases the thickness of the overlayer on the nanoparticle surfaces, although it is too small to be directly observed in the TEM (even if the effects of electron beam damage on the carbon overlayer could be eliminated). The O 1s photoelectron peak after water exposure shows an increase in a shoulder around 531.5 eV, consistent with an increase in OH^- species attached to the surface. The increase in the peak in the same region after exposure to the NOM solution is likely to also include oxygen species present in the NOM. The time-dependent XRD measurements imply that the presence of NOM on the particle surface slows oxidation of the nanoparticles.

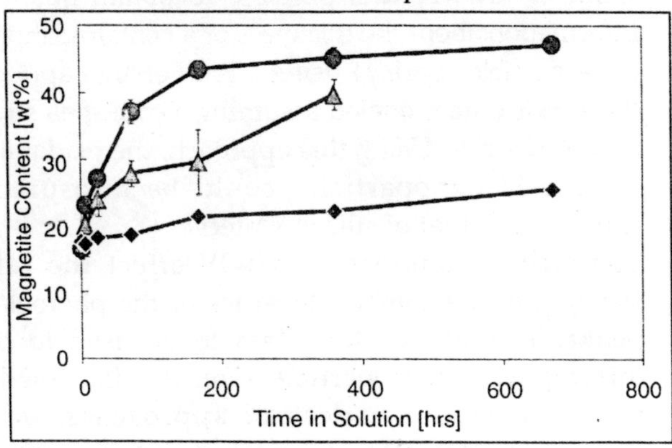

Fig. 6.4

This simple example clearly shows how changes in the environment of nanoparticles and coatings on their surface impact the behaviour of nanoparticles. Many different examples of the use of electron spectroscopy to monitor the presence of contamination, the effectiveness of cleaning processes, or the extent of oxidation have been reported in the literature. The examples highlighted here demonstrate the range of systems that can be examined. XPS and near edge X-ray absorption fine structure (NEXAFS) were used to examine the ability to remove the coatings formed on Rh nanoparticles synthesized in a polyvinylpyrrolidone (PVP) containing solution. The presence of a contamination layer was identified by XPS using standard analysis methods. Of equal importance was the ability to monitor the removal of a contamination layer. In another study AES, due to its higher spatial resolution, was used to monitor contamination removal on a Si substrate supporting an array of nanoparticles. The oxidation state of Si and the presence of O on Pt nanoparticles were also examined. Additional analysis tools used in this study included AFM, scanning electron microscopy (SEM), and XPS. When particle shape is known, it is possible to obtain quantitative information about the thickness of a contamination layer (or a particle coating) using XPS. For this application, the XPS is data modeled assuming a core shell structure of the particles. Using this approach, the oxidation rate for the Si nanoparticles could be measured and compared to that of silicon wafers.

- The size of nanoparticles will affect the relative strength of the signal intensities of the photoelectron peaks. This allows XPS data to be used to obtain information about particle size. As discussed in a recent review, a variety of approaches, initially developed for small metal catalyst particles, have been developed to extract this information. For example, the ratio of photoelectron intensities (from spherical particles of Cu or other metals) having different escape depths can be used to approximate the average particle size. Background signals and photoelectron binding energies of nanoparticles can also be used to determine sizes of spherical particles. The use of energy loss electrons (usually considered the background signal) for Au particles on a polystyrene substrate provides one specific example. The particle sizes determined by XPS compared well to those determined by electron spectroscopy. Many studies have shown that metal clusters and nanoparticles, particularly when they are supported on nonmetal supports, have XPS binding energies that differ from those of bulk materials. The binding energy (BE) shifts vary with sufficient regularity that Gonzalez-Elipe *et al.* proposed BE peak position could be a useful way to determine particle size. Nosova *et al.* determined the relations among the BE, particle size, and turnover number for hydrogenation of vinylacetylene by Pt particles on a variety of supports. In each of these approaches, an average particle size is determined, and the particle shape is assumed. Whenever possible, it may be useful to verify the actual shape by electron microscopy (TEM or SEM) or, possibly, SPM.

- The resolution available with the electron beam allows AES to be used to collect information about individual nanoparticles or nanoparticle arrays. Liang *et al.* have been interested in the formation of Cu_2O nanodots for possible chemical or photochemical applications. AES was used to examine the nature of the nanodots formed on a $SrTiO_2$ substrate after deposition using oxygen plasma assisted molecular beam epitaxy. A secondary electron image of the nanodots, along with AES maps for Cu, Ti, and O, are shown in Figure 6.5. One objective of the AES analysis was to determine if a Cu_2O wetting layer was observed between the Cu_2O nanodots.

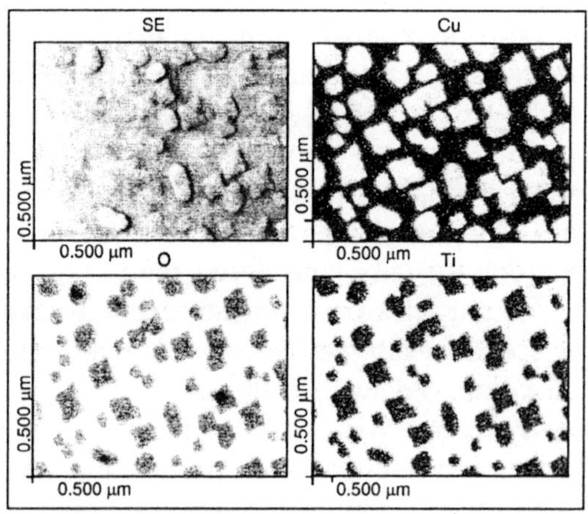

Fig. 6.5

Such a wetting layer was not observed. The secondary electron (SE) image and AES maps indicate that nanodots can be formed with differing shapes. This process has been examined as a function of the amount of material deposited, and AFM measurements will be shown in a later section. AES combined with SEM and TEM has also been used to examine the location of a composite of organic-inorganic nanoparticles (COIN) on leukemia cells. The combined analysis approach provided reliable high-resolution information about the nanoparticles and their binding to cell surface antigens. In this example, AES proved to be useful, despite a coating applied to minimize charging on the biological surface. AES has also been used to measure the concentrations of Au nanoparticles grown within a polyelectrolyte matrix. The need here was to identify the location and amount of gold retained on the polyelectrolyte brush surface. TEM, AFM, and X-ray reflectivity (XRR) measurements completed the tool set to obtain the detailed understanding of this complex nanostructured material.

- Determining the nature and distribution of active sites on nanostructured surfaces is an important challenge with relevance to catalysis and, possibly, particle toxicity. XPS has been used to identify and quantify the presence and distribution of Brønsted and Lewis acid sites for ZSM-5 zeolites. The method involved the deconvolution of N 1s XPS features for pyridine chemisorbed on the zeolite. Three different N states were identified and assigned to Lewis sites and weak and strong Brønsted acid sites. Comparison of the XPS data with IR spectroscopic data suggested that XPS could be used to both identify and quantify the nature of surface acidity.
- In addition to the composition and chemistry, it has been possible to use XPS in combination with sample substrate biasing and controlled electron flood gun voltage to obtain electrical information from films and particles. By biasing a collection of nanoparticles while conducting XPS, it is possible to learn aspects of the electrical properties of Au/silica (core/shell) nanoparticles, particularly for particles embedded in a layer.

ION-BASED SURFACE ANALYSIS METHODS

Ion beams can be used in a variety of ways to obtain information about the nature of nanoparticles. One of the primary uses of TOF-SIMS is to extract molecular information about the functional groups and, possibly, molecular orientation of molecular coatings on particles surfaces. During SIMS measurements primary ion beams of Ga^+, Ar^+, O_2^+, Cs^+, C_{60}^+, Au^+, Bi^+, or other atomic, molecular, or cluster ions with energies between 3 and 20 keV strike the sample surface and result in the ejection of secondary ions. To extract surface molecular information, TOF-SIMS is used in a "static" mode that involves a low density and low total dose of ions such that the surface damage and alteration is minimized. Both atomic and molecular secondary ions are used to extract the surface information.

Also known as Ion Scattering Spectrometry (ISS), LEIS is a well-established, but not widely used method. However, recent developments have made it particularly useful because of the high sensitivity to the outermost atomic layers of a sample. In LEIS, a low energy ion beam (typically an inert gas ion) is scattered off of the surface.

The amount of energy lost by the ion during this scattering process is used to determine the identity of the elements present in the outermost surface of the material under analysis. Here, low energy means ion energies less than a few thousand electron volts (which is low relative to the million volts used in Rutherford backscattering spectroscopy). The high surface sensitivity (near surface depth resolution) makes LEIS a particularly useful tool for surface enrichment measurements of catalysts or a wide variety of other materials.

Like the electron spectroscopies, SIMS is useful for obtaining molecular information about surface layers, functional groups added to the surface, and contamination. Three differences between the electron spectroscopies and SIMS are the higher detection sensitivity of TOF-SIMS for some species, the complicated and nonlinear signal dependence of TOF-SIMS signals (which may complicate quantitative analysis), and the changes generated at the surface by ion sputtering. The high sensitivity of TOF-SIMS for some functional groups has been usefully applied in many ways.

For example, it has been used to examine peptides conjugated to gold nanoparticles as part of a protein kinase assay and to examine multilayer plasma deposited organic coatings on alumina nanoparticles. For relatively large nanoparticles produced during welding, SIMS with sputter profiling has been used to examine the complex layers that forms on these particles. It must be noted that the sputtering of nanoparticles and particles in general may be significantly different from that of thin films or bulk materials. Therefore, sputter depth profiling is not likely the optimum approach for getting layer or depth information about nanoparticles. In addition to understanding surface coatings or functional groups, SIMS has proven to be equally useful for examining the basic composition of nanoparticles, both as they are being processed inside the TOF-SIMS vacuum system and after they have been introduced into the vacuum system for analysis.

In situ thermo-TOF-SIMS was used to examine the thermal decomposition of zinc acetate dehydrate during nanoparticle formation within a SIMS system. SIMS, in combination with TEM and *in-situ* optical transmission spectroscopy, has been used to study the composition and plasmon resonance of unique ZrN nanoparticles produced by laser ablation/evaporation and adiabatic expansion from zirconium nitride powder targets. TOF-SIMS has also been used to characterize the composition and oxidation state of spark-generated nanoparticles made from pairs of Ir-Ir and Ir-C electrodes. From detailed analysis of the SIMS data, it was also possible to obtain information about particle size.

LEIS seemed to be going out of favour until new instrumentation developments significantly enhanced the sensitivity of the method, making it useful for characterizing several nanoparticle properties. It can sometimes be used to determine information about surface layers that might be missed by other surface sensitive techniques.

A particularly interesting example is the use of LEIS to examine the outer surfaces of poly(propylene imine) dendrimers. Although the different generations of dendrimer have the same overall C/N ratio, the outer C/N surface ratio changes as the dendrimers age. By combining the high surface sensitivity of LEIS with the "near" surface sensitivity of XPS, it was possible to show that dendrimer aging changes primarily the outermost atomic layers.

As with XPS, LEIS does not have the resolution to examine individual particles, but can be used to measure the average size of particles and to look at changes in particle size as a function of time. It has been used, for example, to determine metal segregation and the average cluster size of Pt/Rh/CeO_2/ã-Al_2O_3 supported catalysts. The cluster size measurement has been applied to atomically dispersed metals in a nanostructured catalyst and can be used to determine cluster size for particles up to 10 nm.

SCANNING PROBE MICROSCOPIES

STM and AFM are powerful techniques that have enabled major advances in nanotechnology. The development of these "sharp" tip based probes has made it possible for almost any laboratory around the world to examine surfaces of many types of materials with spatial resolutions approaching 1 nm for electrical properties (STM) or sample topography (AFM). AFM can provide 3-D imaging/visualization of nanoparticles distributed on a flat surface.

It can provide qualitative and/or quantitative information about physical properties of nanoparticles including size, morphology, surface texture, and roughness. However, the influence of the AFM tip size and shape on the acquired images must be properly accounted and corrected for. Collection of a variety of images can be used to extract information about particle size distributions and volumes. AFM can be conducted in vacuum, ambient conditions, liquids, or other environments.

The application of SPM methods to nanoparticles takes many different forms. Many, but by no means all, applications involve examining nanoparticles on flat surfaces where many different types of information can be extracted including size and size distribution, as well as shape (including facets). The shapes of the Cu_2O nanodots formed on a $SrTiO_2$ and previously described have been examined by AFM (Figure 6.6). The figure 6.6 shows the size and shape of these nanodots that form on the substrate as a function of the total amount of Cu_2O deposited on the substrate.

Somewhat uncommon applications of SPM methods have proven to have high value and two examples are noted here. By attaching a nanoparticle to a scanning probe tip, it is possible to measure the interactions of individual nanoparticles with a flat surface (possibly chemically functionalized) or to other nanoparticles attached to a flat surface.

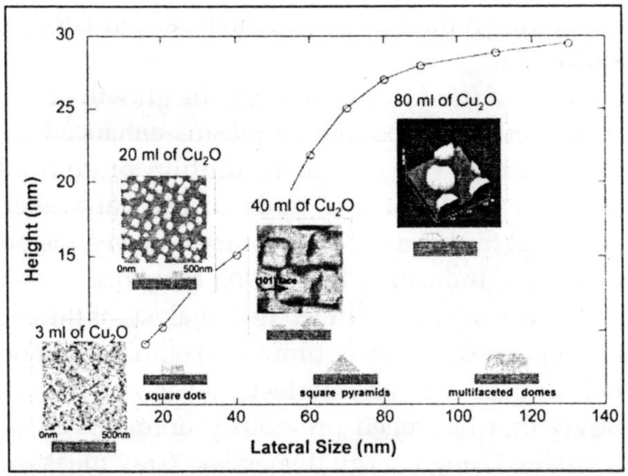

Fig. 6.6

Such measurements provide information about interaction forces and enable these forces to be examined in different environments. In some circumstances, it is useful to determine the roughness of nanoparticle surfaces where many standard methods of measuring surface roughness cannot be applied. AFM in combination with TEM has been used to investigate the impact of synthesis processes and particle size on the surface roughness of ceria nanoparticles.

CHARACTERIZATION OF CARBON NANOTUBES (CNTS)

A wide range of tools have been applied to characterize and understand various forms of CNTs, including single-wall carbon nanotubes (SWCNTs or SWNTs) and multi-walled CNTs (MWCNTs), as they are synthesized, functionalized, and applied. No attempt will be made here to summarize all appropriate methods. Electron microscopy and optical methods have proven to be highly valuable. However, in many circumstances, surface analysis tools, also in combination with a variety of other methods, have proven to be important for obtaining critical information. Examples include the use of surface tools to optimize CNT growth conditions, study the effects of doping on the electronic structure, and verify surface functionalization.

AES has been used to characterize the growth of vertically aligned CNTs on a Si substrate by plasma-enhanced chemical vapour deposition. The spatial resolution of AES allowed examination of the head and body of the nanotubes. AES measurements at the head of the nanotubes showed the presence of the Ni catalyst, indicating that the Ni remained at the top of the nanotubes during growth. Detailed analysis of the substrate allowed the deposition to be optimized to eliminate amorphous carbon byproducts. A 5 nm interfacial layer was identified on the substrate that had been previously unidentified because previous studies using energy dispersive X-ray measurements did not have the surface sensitivity to see the layer.

Many studies of CNTs involve efforts to alter their surface properties by surface functionalization or to alter the overall properties of the nanotubes by doping the particles in some way. Surface tools can provide information about the presence of the atoms or molecules added to the CNTs and the extent of charge transfer between the CNTs and the added material. The noncovalent surface functionalization of SWCNTs using succinimidyl ester for attachment of proteins was previously described. A review by Daniel *et al.*summarizes the efforts to functionalize CNT surfaces with DNA and the approaches towards characterization. The purpose of DNA functionalization has been both to create surfaces useful for sensing and as a method to direct the assembly of nanotubes. XPS and gas chromatography-mass spectrometry (GC-MS) thermal analysis were used to examine some details of CNTs functionalized by solvent-free and aqueous-based arenediazonium. Both methods suggested that only the arcne group was retained on the CNTs after the process.

Another approach to functionalization involves the sulfuric/nitric acid oxidation of CNTs or other graphene-containing materials. A systematic, time-dependent, surface-sensitive study used XPS, Fourier transform infrared (FTIR), and Raman spectroscopies to examine in real time the evolution of the functionalization of carbon fibres sonicated in a mixture of the concentrated acids. The study showed that oxidation occurred after the acid attack created sites that were subsequently oxidized. Smith *et al.* used XPS to characterize the surfaces oxidized by the acid process as part of a study of the impact of these surface oxides on the colloidal stability of MWCNTs.

Efforts to alter the electronic structure of CNTs have involved both adding atoms within the structure of the CNTs and by attaching molecules to the surface likely to be involved in charge transfer. Tetracyano-p-quinodimethane (TCNQ) was shown by core level and valence band XPS, in combination with NEXAFS, to deplete the ð-band density of states of SWNTs because of transfer to the TCNQ. Another method of doping to change the electronic structure has involved low energy implantation of N. XPS has been used to examine both the amount of N implanted and the electronic structure of the implanted SWCNTs.

Charge transfer between Au nanoclusters and as synthesized and plasma oxidized CNTs have been examined by TEM and XPS. Core-level and valence band measurements showed little charge transfer between the Au clusters and MWCNTs. The authors of this paper note that the uncleanliness of the surface, the presence of amorphous phases, and agglomeration of as grown CNTs appear to be constraints to CNT applications

CONSIDERATIONS

Some general issues associated with nanoparticles are useful to consider when analyzing nanoparticles. We highlight here some of the challenges already briefly noted. Billinge and Levin have described the challenges for three dimensional structural characterization of nanoparticles. The level of detail that scientists, engineers, and manufacturers would like for nanostructured materials exceeds that currently possible, particularly on a routine basis. In particular, new methods that involve novel integration of information from more than one method and the application of theoretical analysis are needed. The following list of challenges and opportunities outlines some of these generic needs.

A variety of studies have shown that nanoparticles can be *unstable with respect to the environment*. The inherently lower level of stability for many types of nanostructured materials significantly increases the attention that analysts need to pay to the impacts that an analysis probe, environmental conditions, and time can have on the materials analyses. For example, the sorption of water onto the surface of ZnS nanoparticles changes the particle structure. The energies associated with many of the probes used for particle analysis equal or exceed those needed to transform the particles in a variety of ways. We have observed significant time-dependent reactivity associated with Fe metal-core oxide-shell nanoparticles when placed in water. These issues impact the care needed for sample handling, the time and nature of the analyses, and the need for comparison of results from a variety of methods.

The significant fraction of atoms or molecules associated with surfaces and interfaces increases the potential impact of surface impurities, surface enrichment or depletion, and *surface contamination* on nanoparticle properties and complicates accurate measurements. Surface coatings and impurities can significantly alter many aspects of nanoparticle behaviour. The deliberate application of coatings has been used to control particle shape during growth, and the use of coatings on nanoparticles can be used to control particle spacing and the resulting properties of nanoparticle composites. Because of *capillary* and *sorption* effects, the high surface area present in collections of nanoparticles may retain solvent in circumstances that can surprise researchers. Even using surface tools, it can sometimes be difficult to characterize the nature of the actual nanoparticle surfaces.

Surface tools are highly valuable in measuring surface coatings and contamination layers, but it can be important to combine shape or other data obtained by *complementary methods* to increase the value of the information that can be obtained.

Shard *et al.* for example, show how XPS can be used to accurately determine coating thickness when particle size and shape are known.

Because the *physical properties* of nanoparticles can change *with size and the environment*, some of the assumptions made in the analysis (*e.g.*, electron inelastic mean free paths) may be inaccurate. Samples are a polymorph of bulk materials, which may impact the accuracy of electron spectroscopy measurements on nanoparticles. Ion beams will damage nanoparticles at rates significantly higher than typical for bulk materials for thin films. Sputtering of nanoparticles can be 10 times faster than for "bulk" films. Such behaviour can impact SIMS measurements on nanoparticles. Some of these issues are identified as specific technique concerns in a following section.

For the reasons already stated, the preparation, processing, and mounting of nanoparticles requires particular care and attention. A measurement made without reference to *sample history* and *previous processing* will often provide misleading or incorrect information. These effects also place a high value on measurements that can be made without removing materials from the environment of interest. *In situ*(and *real time*) analysis methods provide important new levels of information useful for advancing nanotechnology.

SAMPLE MOUNTING ISSUES

The heightened concerns about contamination, impurities, and environmental effects necessarily raise issues and concerns about the handling and mounting of specimens for analysis. For nanoparticles, the issues associated with sample mounting are similar to those associated with mounting any sample for surface analysis, and a variety of guides have been developed. For the set of surface techniques discussed here, the nanoparticles will normally be supported on a substrate and examined either individually (SPM and, sometimes, AES and SIMS) or as a collection of particles. Confirming the nanoparticles are immobilized without introducing surface contamination may be achieved using a variety of methods. Examples include the use of vacuum compatible tape (after checking for contamination issues), growth on a substrate, solution deposition onto a substrate, or dipping a support into solution.

Different preparation and analysis methods produce a variety of particle distributions as shown in Figure 6.7. Some nanoparticle properties and measurement results can be impacted by agglomeration or aggregation or interactions of nanoparticles with the supporting substrate and/or with other nanoparticles (proximity effects.

Such effects include:

- Buildup of charge during XPS measurements of metal clusters supported on insulating substrate,
- Coupling of plasmon modes in metal nanoparticles within close proximity,
- Coupling of quantum states,
- Impact of particle spacing on electronic and magnetic properties of composite, and
- Effect of "buffer layers" on optical properties of silicon nanocrystal superlattices.

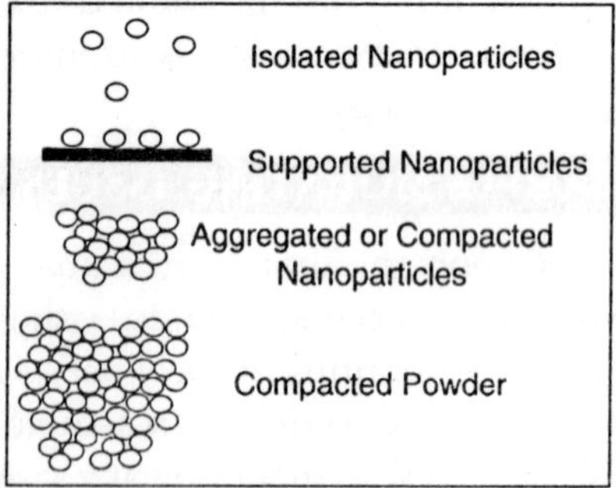

Fig. 6.7

ANALYSIS CONSIDERATIONS FOR SPECIFIC METHODS

In this section, we consider some of the nanoparticle analysis considerations relevant to the specific methods discussed in this paper. The topics are not comprehensive, but suggest both areas to be addressed and some of the solutions currently used to address them.

XPS CHARACTERIZATION OF NANOPARTICLES

XPS has been used to analyse nanolayers almost since its inception. One of the major application areas has been the characterization of complex catalysts particles, which often included supported metal nanoparticles as the active components. A wide variety of nanostructures are now routinely analysed using XPS. While most XPS analysis is conducted assuming that the sample has a uniform flat surface layer, the shape and structure of nanoparticles can play an important role in accurate interpretation of XPS (and other) data of any nanostructured material. As suggested by Figure 6.7, such materials do not present a uniform surface for analysis. Several analytical methods can be used to extract important information about nanoparticles and nanostructured materials by moving beyond the uniform layer analysis method. The appropriate approach also depends on the density and mounting of the particles. Often modeling of the experimental data is required to extract information not available from the standard simple analysis. Examples are provided as follows.

LOW DENSITY OF SUBSTRATE SUPPORTED PARTICLES

Although the size of nanoparticles can be determine by both peak ratio and BE measurements, as noted earlier, the peak ratio method applies most readily when the particles are present at low density (not overlapping) on a surface. An interesting feature of this approach is that the size of the particles can be determined even if they are too small to be observed by electron microscopy. However, if the particles are covered by contamination or buried within a substrate photoelectron peak ratios will be influenced by the covering layers and this method is not generally valid (or may require an overlayer correction). With the assumption or verification that the nanoparticles are primarily on the surface, the kinetic ratio method appears to apply equally well to rough as well as ideally flat surfaces.

In the ideal circumstance that nanoparticles are distributed as single layer on a flat surface, it is possible to use angle-resolved XPS to obtain information about particle size, size distribution, and separation. It is also possible to examine the presence and even thickness of coatings, including corrosion or oxidation layers (these can be considered core shell type particles). For this type of analysis, sample preparation, the mounting method and the nature of the substrate are important because particles may interact either chemically or respond physically to the nature of the substrate. Also, it is desirable to minimize possible overlap of peaks between particles and the substrate.

Variable excitation energy XPS (ERXPS), available at synchrotrons, offers a relatively new and increasingly powerful alternative to obtaining depth information from powdered samples (such as nanoparticles) and other real surfaces that are not ideally smooth. Merzlikin *et al.* have shown that ERXPS can be used to collect useful coating information even for very rough surfaces (including a collection of nanoparticles as described below). Other work by Merzlikin directly showed the ability to use the ERXPS approach to examine core-shell nanoparticles.

AGGLOMERATES OF PARTICLES

It is frequently not possible to distribute particles at low density, and, in some circumstances, a larger collection of particles is needed to obtain sufficient signal. In a now classic study, Fulghum and Linton compared analysis methods to determine quantitatively the coverage of a specific surface adsorbate on the surfaces of particles in a collection of particles. They showed that XPS can be used to obtain accurate surface coverage even on the surface of a powder sample. Castner and Campbell have also presented a detailed method for XPS analysis of nanoparticle agglomerates.

For nanoparticles, knowledge about the surface coating and/or contamination may be central to their application. The approach of Shard *et al* may be applied for both single layers of particles or agglomerates if the particle size is known. Agglomerates of particles present what is effectively a very rough surface for analysis.

The so called "magic" angle (analyser geometry) presents another approach to data collection and analysis that can be used to estimate layer coverage, with appropriate consideration of the accuracy limits.

SIZE AND CURVATURE EFFECTS

The signal strength from a coating relative to that of the substrate (or particle core) can vary with particle size because of the electron path lengths and surface curvature. Data collected on different sized Au nanoparticles coated with C16 COOH thiol molecules in a self assembled monolayer (SAM) shown in Figure 6.8 provide a demonstration of this effect. Although the SAM layer thickness is nearly constant, the ratio of the C signal from the SAM to the Au signal from the core particle changes as a function of particle size.

As the particle gets smaller, the collection of particles contains an increasing percentage of SAM material. Surface curvature effects impact relative peak intensities of overlayers and substrates (cores) for all sizes of particles, but when the particle size becomes smaller than the electron escape depth it also becomes possible to detect photoelectrons from both the top and bottom of the particles.

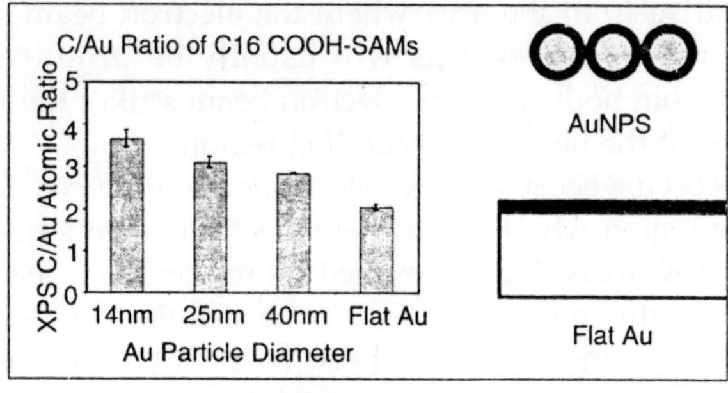

Fig. 6.8

BINDING ENERGY AND PEAK WIDTH CHANGES

The size of particles (and interactions with the substrate) can impact the binding energy and peak width of photoelectron peaks, change the valence band peak shape, and alter the Auger parameter. These effects sometimes can be used to determine particle sizes, as noted above.

The BE method for size determination may be less impacted by surface contamination than the peak ratio method and has been used as a measure of Au nanoparticle size for particles grown within polyaniline films.

However, not all size dependent BE shifts are confined to nanoparticles—the binding energy of SiO_2 on Si varies with the film thickness due to interactions with the substrate. It can be important to remember that not all shifts in photoelectron binding energies associated with nanoparticles should be interpreted as chemical state changes.

AES ANALYSIS OF NANOPARTICLES

Many of the considerations described for XPS apply to AES as well because they relate to the impacts of particle size and distribution on the low energy electrons detected by the AES or XPS process. However, because AES involves electron excitation, issues of electron backscattering and transmission through a particle provide some additional areas to consider. There is a general tendency of analysts to look at a secondary electron image and assume that all of the AES signals from a feature arise from the area where the electron beam strikes. However, Auger electrons will usually be produced and detected from both where to electron beam strikes the sample and some of the nearby surrounding region.

This occurs because Auger electrons are produced not only from the region where an incident electron beam strikes, but from nearby areas that are excited by the primary beam as it penetrates into and is scattered by the sample. This is a well-known phenomenon, and there is an ISO standard for determining the analysis area and lateral resolution for AES and XPS analysis.

TOF-SIMS ANALYSIS OF NANOPARTICLES

TOF-SIMS analysis involves the bombardment of a substrate with ions that have energies on the order of tens of keV (which is comparable to the cohesive energy of a nm-sized cluster and typical penetration lengths (for monatomic ions on flat uniform substrates) up to tens of nm.

The affected area from a single collision is also on the order of one to tens of nm. Thus, the interaction area and energy of the primary ion is of the same order of magnitude as the size of a nanoparticle. Simulations indicate this may lead to a number of potential issues, challenges, and artifacts.

The primary effect that is seen through both simulation and experiment is that the sputter yield/rate may increase several fold. Simulations show this to be due to recoil from the "back" and escape from the "sides" of the nanoparticle. It is also useful to recognize that with sputter time, the shape of particles will change.

The enhanced sputtering of nanoparticles has a potential impact on the information that can be obtained from organic or other coatings on these particles. The static limit is identified as the ion doses below which the molecular information is typically maintained, usually quoted as 10^{12}–10^{13} ions/cm^2. At these doses, on the order of 0.1–1% of the surface sites have been subject to direct impact.

Because TOF-SIMS is often used to characterize functional groups associated with coatings on nanoparticle surfaces, it is desirable to stay well below the static limit and to know that this limit may be significantly different for nanoparticles relative to flat surfaces.

To determine the static limit for nanoparticles, an experiment was devised whereby alcohol-functionalized titanium dioxide nanoparticles (TiO_2 nps) were covalently attached to a silicon substrate covered with native oxide and an isocyanate self-assembled monolayer. The alcohol reacts with the isocyanate via standard urethane chemistry to yield a monolayer of TiO_2 nps. In practice, this substrate also has a number of large aggregates.

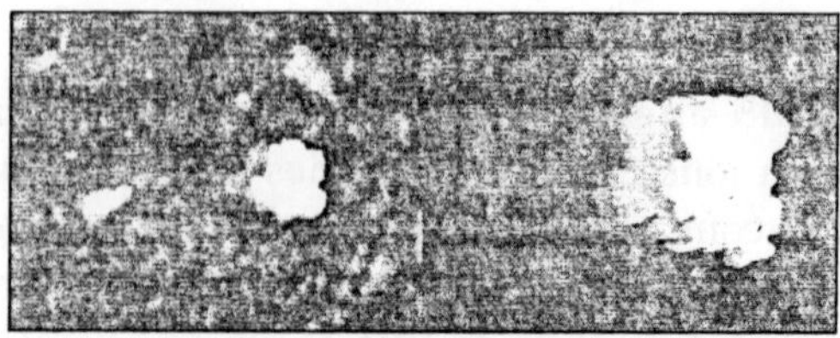

Fig. 6.9

The substrate was sputtered using a PHI TRIFT II TOF-SIMS instrument with 15 kV Ga^+ for a dose ranging from 5×10^{11} ions/cm^2 to 3×10^{16} ions/cm^2. Peak intensity data for $^{48}Ti^+$, $^{48}Ti^{16}O^+$, and $^{8}Ti^{16}O_2^+$ were recorded for doses up to 1×10^{16}ions/cm^2 to measure differential sputtering of O, known to occur widely with metal oxides.

These data are summarized in Figure 6.9. As is evident from the data, differential sputtering starts to occur for doses above 5×10^{12} ions/cm^2 within the conventional measure for the static limit. Other data indicate that the organic coating on the nanoparticles and substrate (whether from the SAM or adventitious carbon) shows some damage at a slightly lower ion dose, but still above 1×10^{12} ions/cm^2. However, this analysis is complicated by the observation that the Ti and TiO_x signals do not disappear when the individual nanoparticles are no longer seen in the SEM. This indicates that the secondary ion signals are arising primarily from the larger aggregates.

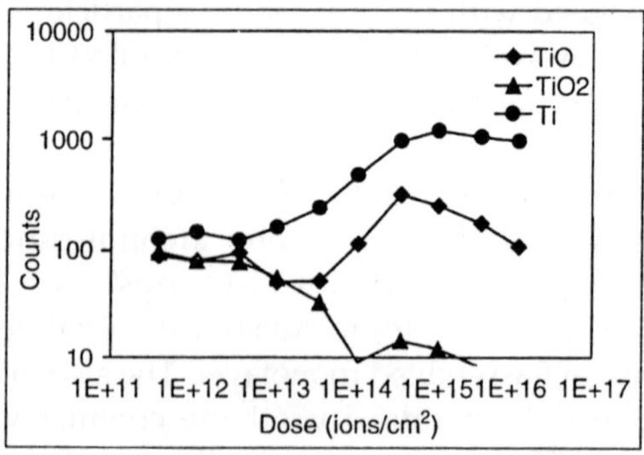

Fig. 6.10

This data indicates that aggregates of particles behave in a similar fashion as bulk materials (*i.e.*, secondary ion (SI) signals are not affected by the nano-size character of the sample). However, in a sample with a mix of aggregates and individual nanoparticles, the signal arises primarily from the aggregates. This may be from poor coupling of the incoming primary ion to the nanoparticle, leading to little energy deposition in the nanoparticle. Substrate signals may dominate over SI signals arising from individual nanoparticles in a sparse monolayer. For cluster beams, this effect may be negligible. The sensitivity of the nanoparticles to perturbations leading to amorphization, loss of secondary structure, reduction, etc., may be influenced by the three-dimensional structure. Because of the angle dependence of sputter rates, sputtering can vary as much as 10 times for different portions of a nano-object.

Sample history and handling may also influence the SIMS analysis of nanostructured materials in a similar manner to other surface analysis methods. One other complicating factor is that the high electric fields used in the extraction region to accelerate the secondary ions into the spectrometer may induce "field-emission" of nanoparticles.

SCANNING PROBE MICROSCOPY ANALYSIS OF NANOPARTICLE

Although a wide variety of information about nanoparticles supported by a flat substrate can be obtained using SPM methods, there are a variety of tip artifacts that can influence the accuracy of the measurements.

7

Ecotoxicology of Manufactured Nanoparticles

INTRODUCTION

The soil root interface (the rhizosphere) plays a vital role in sustaining life in the terrestrial ecosystem in the Earth's Critical Zone (CZ). The CZ is defined as the volume extending from the upper limit of vegetation down to the lower limit of groundwater.

The CZ is the system of coupled chemical, biological, physical, and geological processes operating together to support life at the Earth's surface.

While our understanding of this zone has increased over the last hundred years, further advance requires scientists to cross disciplines and scales to integrate understanding of processes in the CZ, ranging in scale from the mineral-organic matter-organism-water-air interfaces at a molecular level to the globe. Soil is the central organizer in the CZ. The rhizosphere is a hot spot of activity within soils.

The rhizosphere is the 'bottle neck' of the supply of vital elements to sustain ecosystem productivity and integrity and food security.

The rhizosphere is also the 'bottle neck' of the contamination of the terrestrial food chain by inorganic and organic pollutants to endanger human and animal health.

PHYSICAL, CHEMICAL AND BIOLOGICAL INTERACTIONS AT THE SOIL-ROOT-BIOTA INTERFACES

Solid-liquid interface character should vary substantially with lithology, climate, vegetation, below-ground biota, landscape position, time, and anthropogenic activities. The rhizosphere should greatly influence solid-liquid interface character and, thus, the nature and properties of their reactive surfaces. In the rhizosphere, the kinds and concentrations of substrates are different from those in the bulk soil because of root exudation.

This leads to colonization by different populations of bacteria, fungi, protozoa, and nematodes. The plant-microbe interactions result in intense biological processes in the rhizosphere. These interactions, in turn, affect physicochemical reactions in the rhizosphere. Physicochemical properties that can be different in the rhizosphere include: acidity, concentration of complexing biomolecules, redox potential, ionic strength, nutrient status, enzyme activity, bulk density and porosity, and moisture.

These differences in soil physicochemical properties would, in turn, influence biological processes in the rhizosphere. The total rhizosphere environment is governed by an interactive trinity of the soil, the plant, and the organics associated with the roots. The reactions and processes in the rhizosphere can only be understood satisfactorily with interdisciplinary approaches. The impacts of physicochemical and biological interfacial interactions on food security and ecosystem integrity, thus, merit serious attention.

RHIZOSPHERE INTERFACIAL INTERACTIONS AND FOOD SECURITY

The rhizosphere chemistry, biology, and physics and their interactions at the molecular level are of fundamental and practical importance in sustaining food security. The rhizosphere is the "bottle neck" of the supply of nutrients and the ecotoxicological effect of inorganic and organic contaminants to plants. Therefore, the impacts of the rhizosphere on plant health and crop production is a critical issue in feeding the world population. 30% of farmers in developing countries are food-insecure. Some of the most profound and direct impacts of climate change over the next few decades will be on agricultural and food systems.

Therefore, food insecurity is likely to increase under climate change, unless early warning systems and development programmes are used more effectively. Managing world soils for food security and environmental quality is an extremely important mission of soil scientists.

Transform agricultural systems through improved seed, cropping system, fertilizer, land use, and governance, and food security may be attained by all. In this context, rhizosphere management deserves close attention in sustaining and enhancing soil productivity and crop production.

IMPACTS OF RHIZOSPHERE INTERFACIAL INTERACTIONS AND ECOSYSTEM INTEGRITY

Carbon Transformation, Storage, Emission, and Climate Change

The CO_2 emission from the soil to the atmosphere is the primary mechanism of soil C loss. Agricultural practices contribute about 25% of total anthropogenic CO_2 emission. Soil minerals control carbon storage and turnover.

Respiration by plant roots contribute about half of CO_2 emitted from the soil. The rhizosphere significantly controls SOM decomposition.

Given abundant mineral nutrient supply, soil microbes prefer labile root-derived C to SOM-derived C, resulting in a decreased SOM decomposition in the rhizosphere. If mineral nutrients are in short supply, soil microbes prefer nutrient-rich SOM to root-derived C, resulting in increased SOM decomposition in the rhizosphere. Root effects on the rate of SOM decomposition can range from negative 70% to as high as 33% above the unplanted control.

FORMATION AND TRANSFORMATION OFNANOPARTICLES AND ECOSYSTEM HEALTH

Nanoparticles are discrete nanometer (10^{-9} m) scale assemblies of atoms. A significant fraction of atoms are exposed on surfaces rather than contained in the particle interior of nanoparticles (ca. 1-100 nm). The biogeochemical and ecological impacts of nanomaterials are some of the fastest growing areas of research today, with not only vital scientific but also large environmental, economic, and political consequences. Most biominerals generated by microorganisms are nanominerals. Therefore, microbial processes play an important role in nanoparticle formation and fate. Little is known on nanoparticles in the rhizosphere. Formation, transformation, and fate of nanoparticles in the rhizosphere and the impacts on food security and safety and ecosystem health merit increasing attention.

TRANSFORMATION OF METALS, METALLOIDS, AND XENOBIOTICS AND FOOD SAFETY

The interactions at physical-chemical-biological interfaces govern the kinetics and mechanisms of transformation, speciation, transport, bioavailability, toxicity, and fate of metals and metalloids in soil and related environments. Fundamental understanding of soil physical, chemical, and biological interfacial interactions at the atomic and molecular levels is essential to understanding the behaviour of metals and metalloids in the pedosphere and restoring terrestrial ecosystem health on the global scale. We are still at the eve of fully understanding the processes controlling the mobility and bioavailability of metals and metalloids at the soil-root interface. Unraveling the biogeochemistry of trace elements in the rhizosphere and the impact on food safety and ecosystem health is both a challenge and an opportunity for soil and environmental scientists for years to come.

The multiple interactions between microorganisms and soil which take place at the soil-root interface can have a dramatic impact on anthropogenic organic pollutants in the terrestrial environment.The characteristics of the rhizosphere allow for a unique environment which encourages biotic and abiotic interactions in greater magnitude and diversity than in bulk nonvegetated soil. These interactions generally resulted in the decrease in the toxicity of organic contaminants through increased degradation and decreased availability of these contaminants. As the importance of bioavailability has increased in regulatory decision making, a greater understanding of the plant, soil, and managing factors affecting the bioavailability of contaminants is warranted.

Ecotoxicological Problems

Ecotoxicology is defined as "the study of fate and effect of toxic agents in ecosystem". Ecotoxicology research deals with the interactions among organisms, toxic agents (in this case metals and metalloids), and the environment. Long-term ecological effects of metals and metalloids introduced to soils remains to be investigated, as only a few such experiments exist. Even less is known about the adverse long-term effects of metals and metalloids on soil microorganisms

BIODIVERSITY

The functioning and stability of the terrestrial ecosystem are determined by plant biodiversity and species composition. Above and below ground communities can be powerful mutual drivers, with both positive and negative feedback. Above-and belowground components are closely interlinked at the community level, reinforced by a greater degree of specificity between plants and soil organisms. The impact of soil physicochemical and biological interfacial interactions in the rhizosphere on belowground biodiversity remains to be uncovered.

GEOMEDICAL PROBLEM AND HUMAN HEALTH

Geomedicine is the science dealing with the environmental factors that influence the geographical distribution of pathological and nutritional problems relating to human and animal health. Knowledge of soil science is indispensable for solution to many geomedical problems. The effects of soil physical, chemical, and biological interfacial interactions in the rhizosphere on quality of vegetation and the food and feed produced and related geomedical problems warrant in-depth research. The transformation and bioavailability of trace elements are profoundly influenced by soil physical, chemical, and biological interfacial interactions in the rhizosphere. Many trace elements are of concern to animal nutrition and human health.

These include Se, Fe, I, Zn, Cu, Mn, Mo, Cr, F, Co, Si, V, Ni, As, and Mg. It is essential to promote research on the relationship between soil physicochemical-biological interfacial interactions, especially in the rhizosphere and the impacts on the transformation, transport, and toxicity, and fate of trace elements in the terrestrial environment. This would facilitate fundamental understanding of the linkage of trace elements in soils with plant-animal-human-environmental systems and related geomedical problems. This is essential to provide practical solutions to their deficiency and toxicity problems.

SIGNIFICANCE OF RHIZOSPHERIC INTERFACIAL INTERACTIONS IN RISK ASSESSMENT AND RESTORATION OF ECOSYSTEM INTEGRITY

Regulatory-driven risk assessment and management are essential in restoration of ecosystem integrity. Assessing exposure to contaminants in soil environments includes determination of the pathways to human exposure. The cleanup of soils polluted by hazardous contaminants has become a matter of urgent concern. Physical, chemical, and biological interfacial interactions in the rhizosphere play a significant role in the natural remediation and restoration of the terrestrial ecosystem. Boiremediation, phytoremediation, and chemical remediation have been commonly used in land management practices.

Combined biotic and abiotic remediation would enhance remediation efficiency. The impact of physical, chemical, and biological interfacial interactions in the rhizosphere on risk assessment and management of contaminants and restoration of ecosystem integrity merits increasing attention.

CONCLUSIONS AND FUTURE PROSPECTS

Soil is the most diverse ecosystem and most important terrestrial resource in sustaining life in the Earth's CZ. Soil physical, chemical, and biological interfacial interactions in the rhizosphere play a vital role in carbon cycling and climate change, the formation and transformation of environmental nanoparticles, the fate of nutrients, metals, metalloids, and xenobiotics, ecotoxicological problems, biodiversity, and geomedicine. Fundamental understanding of soil physical, chemical, and biological interfacial interactions in the rhizosphere at the molecular level is essential for developing innovative strategies for land resource management to sustain food security and ecosystem integrity. Future research on this extremely challenging and important area of science should be stimulated to sustain and enhance ecosystem productivity, services, and integrity in the Earth's CZ and the human welfare.

PHYSICO-CHEMICAL TRANSFORMATION OF NANOPARTICLES

Engineered metallic nanoparticles can be beneficially used in consumer products, agricultural production and environmental remediation processes, but they can also have adverse effects on humans and their ecosystem. Transfer of these particles between aquatic and solid phases significantly affects their exposure to humans, organisms and plants. It is determined by the kinetics of several physicochemical transformation processes that can occur upon release of the particles into the water phase.

The particles can be adsorbed from the water phase onto surfaces or they can aggregate into larger particles and precipitate, decreasing their mobility and availability in the water phase. On the other hand, they can also be dispersed by natural organic matter and dissolution of material from the particle surface itself into solution can occur. The surface properties of the nanoparticles are known to be one of the most important factors that govern their stability and mobility as colloidal suspensions, or their adsorption or aggregation and deposition in aquatic systems. They are mainly dependent on parameters such as temperature, ionic strength, pH, hardness, particle concentration and size, etc. In addition, occurring redox reactions and/or association of nanoparticles with natural organic matter or surfactants added to maintain the stability of colloidal suspensions will further increase the complexity of interactions.

Accordingly, particles released into different types of aquatic environments (*e.g.*, varying hardness, salinity and redox potential) are expected to behave in various ways, which in turn leads to different exposure of humans, organisms and plants. In addition, coagulated, precipitated or adsorbed nanoparticles could be transformed and/or remobilised on medium or longer term when environmental conditions change (*e.g.*, pH, redox potential, hardness, organic matter contents and salinity).

Further sustainable development of nanotechnology thus needs the ability to predict the physicochemical fate of engineered metallic nanoparticles released into aquatic environments. Therefore, kinetics of changes in occurrence of metallic nanoparticles as affected by characteristics of the aquatic medium are currently being studied.

MECHANISMS OF NANOPARTICLE TOXICITY IN THE ENVIRONMENT

Gene therapy can be broadly defined as the introduction of genetic material into a cell for either the suppression of gene expression or the production of a needed protein. Because the eye has well defined anatomy, immune privilege, and accessibility, it is a promising candidate organ to benefit from gene therapy. The ocular surface is covered by two protective mucosal epithelia, the cornea and conjunctiva. These epithelia are in direct contact with the tear film and act as barriers for topically administered drugs. Along with physiologic mechanisms such as blinking and tear clearance, the epithelia limit the efficient penetration of drugs and DNA into the eye.

To achieve efficient delivery of DNA to mucosal cell nuclei, several barriers must be overcome. Among the different strategies explored to improve mucosal delivery, one of the most promising is the use of mucoadhesive nanoparticles (NPs) that are capable of interacting with the ocular mucosa. This interaction increases drug residence time and promotes its transport across the ocular barriers. Our group has developed NPs consisting of bioadhesive and biocompatible polysaccharides, such as chitosan (CS) and hyaluronic acid (HA), intended for gene delivery to the ocular surface. CS is a non-toxic and biocompatible cationic polysaccharide with several applications for the administration of drugs and genes. CS NPs interact and remain associated with the ocular mucosa for extended periods of time, increasing the delivery to external ocular tissues, and providing long-term drug retention. In contrast, HA is an acidic mucopolysaccharide distributed widely in the eye. It has been used for the preparation of microparticles and as a coating material for preformed liposomes, NPs, and plasmid DNA (pDNA) complexes.

In previous studies by our group, we have shown that NPs of HA and oligomers of CS (HA-CSO NPs) have the ability to associate with significant amounts of plasmid pDNA, enter cells, and efficiently deliver the pDNA. In rabbits, these NPs entered conjunctival and corneal epithelial cells without causing ocular discomfort or irritation and without significant effects on tissue morphology and functionality or tear production or drainage.

Improving gene delivery requires developing an understanding of the cellular uptake mechanisms, intracellular stability, and bioavailability of the therapeutic DNA.

Five major cell uptake mechanism are distinguished:

- Macropinocytosis;
- Phagocytosis;
- Clathrin-dependent endocytosis;
- Caveolin-mediated endocytosis; and
- Clathrin- and caveolin-independent pathways.

Macropinocytosis and phagocytosis are actin-dependent endocytic mechanisms mainly used by cells to internalize large amounts of fluids and growth factors (macropinocytosis) or solid particles (phagocytosis).

Endocytosis mediated by clathrin and caveolins comprises multiple mechanisms that allow cells to internalize macromolecules and particles into transport vesicles derived from the plasma membrane. Clathrin-dependent endocytosis enables cargo bound to specific membrane-bound receptors to be internalized. Caveolins are the main protein component of caveolae, flask-shaped invaginations of the plasma membrane that participate in macromolecule internalization. Finally, clathrin- and caveolin-independent pathways transport cargo to the glycosylphosphatidylinositol-anchored-protein-enriched early endosomal compartment.

The presence of HA in the HA-CSO NPs may facilitate NP cellular uptake by receptor-mediated endocytosis. HA is biocompatible, biodegradable, and mucoadhesive, and affects several cellular processes. For instance, HA promotes corneal wound healing through regulation of epithelial cell regeneration and migration. This regulation is achieved by binding of HA to two receptors, CD44 and the receptor for hyaluronan-mediated motility (RHAMM) located in ocular surface epithelia. These receptor-mediated processes may be implicated in the cellular uptake of HA-CSO NPs.

In addition to internalization mechanisms, the intracellular trafficking of DNA-loaded NPs and bioavailability of the transported DNA within the cell are critical elements to study when a new drug carrier is proposed. Among these issues, the degradation of the NPs in the lysosomes following their internalization is a key point. Lysosomes are the terminal degradative compartment of certain endocytic pathways and may negate effective drug targeting, in this case of pDNA, to the nucleus.

We propose that HA-CSO NPs may serve as an effective gene delivery system for ocular surface disorders. Therefore, the goal of the present work was to study the intracellular trafficking of HA-CSO NPs loaded with a model pDNA. We aimed to determine which HA-CSO NP internalization pathways are used by corneal and conjunctival cells and whether or not pDNA reaches the cell nucleus.

METHODS

MATERIALS

Plastic culture ware was obtained from Nunc. DMEM/F12 culture medium and other cell culture reagents were from Invitrogen-Gibco. Reagents 2,3-bis[2-methoxy-4-nitro-5sulfophenyl]-2H-tetrazolium-5-carboxyanilide inner salt (XTT), sodium azide, colchicine, chlorpromazine, and filipin were purchased from Sigma-Aldrich Corp.

For immunfluorescence studies, we used anticaveolin-1 and anti-clathrin (Affinity Bioreagents, Rockford, IL) primary monoclonal antibodies (Abs) and goat anti-mouse IgG F(ab′)2 PE-Cy5 as a secondary antibody. The Vibrant Phagocytosis assay kit and LysoSensor reagent were obtained from Molecular Probes.

PREPARATION OF HA-CSO NPS

HA-CSO NPs were made of the polysaccharides HA and CSOs by a slightly modified ionotropic gelation technique. Briefly, 375 μl of HA solution (0.73 mg/ml), mixed with 50 μl of the crosslinker tripolyphosphate (TPP, 0.59 mg/ml), was added over 750 μl of CSO solution (0.625 mg/ml) with magnetic stirring. HA was previously labeled with fluoresceinamine (fl-HA). CSOs were obtained by sodium nitrite degradation as previously described by Janes *et al*. Briefly, 200 μl of 0.1 M $NaNO_2$ were added to 4 ml of a 10 mg/ml chitosan solution. The reaction was left overnight to complete the degradation, and oligomers of approximately 10–12 kDa were recovered by freeze-drying. HA-CSO NPs were loaded with a model pDNA encoding secreted alkaline phosphatase (SEAP) by incorporating the pDNA in the HA/TPP. The theoretical loading was set at 1% (W/W).

PHYSICOCHEMICAL CHARACTERIZATION AND PDNA LOADING OF HA-CSO NPS

Mean particle size and size distribution (polidispersity) were measured by photon correlation spectroscopy. Also, Z-potential values were obtained by laser Doppler anemometry. Stability of NPs upon storage at 4 °C for up to two weeks was evaluated by periodically measuring the size and Z-potential.

CELL LINES AND CULTURE CONDITIONS

Three different cell lines were used. The IOBA-NHC cell line is a nontransfected, spontaneously immortalized conjunctival epithelial cell line used at passages 71 to 87. Cells were grown in DMEM/F-12 supplemented with 10% fetal bovine serum (FBS), 5,000 U/ml penicillin, 5 mg/ml streptomycin, 2.5 μg/ml fungizone, 2 ng/ml human epidermal growth factor (EGF), 1 μg/ml bovine insulin, 0.1 μg/ml cholera toxin, and 0.5 μg/ml hydrocortisone.

The HCE cell line is a SV40-immortalized human corneal epithelial cell line. Cells from passages 42 to 52 were cultured in DMEM/F-12 supplemented with 15% FBS, 100 U/ml penicillin, 0.1 mg/ml streptomycin, 10 ng/ml EGF, 0.5% DMSO, 5 μg/ml insulin, and 0.1 μg/ml cholera toxin.

The mouse macrophage RAW264.7 cell line was cultured in RPMI 1640 supplemented with 10% FBS, 5% L-glutamine, 5% penicillin, and 5% streptomycin. It was used as a control in the phagocytosis assay.

All cell lines were cultured at 37 °C in a 5% CO_2%–95% air atmosphere. Media were changed every other day, and daily observations were made by phase contrast microscopy.

CELL VIABILITY ASSAY

Viability of 20 μg/ml HA-CSO NP-exposed cells was measured using the XTT toxicity test. Cells were seeded onto 96-well plates (2×10^5 cells/well) and grown until 75% confluence. Culture medium was replaced with fresh phenol red-free RPMI, 48 h after NP incubation. Then XTT solution was added and cells incubated at 37 °C for 15 h. Controls included cells alone and cells exposed to 0.5% benzalkonium chloride (BKC), which induces a significant decrease of ocular cell viability in concentrations greater that 0.05%. Cell viability was calculated as a percentage with regard to control cells. Each test was repeated four times in triplicate.

UPTAKE AND TRAFFICKING EXPERIMENTS

Corneal and conjunctival cell lines were seeded in eight-well multichamber Permanox slides (5×10^5 cells/well) and 24-multiwell plates (8×10^5 cells/well). When the confluent state was reached, cells were washed out in supplement-free culture medium for 1 h, and 100 μl of 20 μg/ml HA-CSO NPs were then added. After 1 h incubation, the cells were washed three times with phosphate buffered saline (PBS) with 0.27% glucose, and fresh medium was added. HA-CSO NP uptake was monitored after 1, 6, 24, and 48 h by fluorescence microscopy and by fluorometry. Each assay was performed four times in triplicate.

For fluorescence microscopy, living cells were viewed under an inverted fluorescence microscope. Cell nuclei were counterstained by Hoescht dye. To assure the intracellular location of the NPs, vertical spatial images were generated by Z-scans in 1 μm steps for 20 images. For fluorometry, the cells were washed to remove any extracellular NPs, then frozen, thawed, and disrupted with 100 μl ice-cold radioimmuno-precipitation assay (RIPA) buffer (10 mM Tris-HCl [pH 7.4], 150 mM NaCl, 1% deoxycholic acid, 1% Triton X-100, 0.1% SDS, and 1 mM EDTA). Fluorescence in cell lysates was measured in a SpectraMAX®M5 multidetection microplate reader at an excitation wavelength of 490 nm and an emission wavelength of 520 nm. The amount of NPs taken up by the cells was expressed in arbitrary units of fluorescence intensity. Negative controls included cell incubation with Hank's balanced salt solution instead of NPs.

To explore different mechanisms of NP interaction with epithelial cells and trafficking across the plasma membrane, uptake of HA-CSO NPs was studied under seven different blocking conditions. In all cases, cells were incubated with the different inhibitors for 30 min before the addition of the NPs, and then co-incubated with the NPs for 1 h.

The phagocytic capacity of both the corneal and conjunctival cell lines was studied to exclude phagocytosis as a HA-CSO NP uptake pathway. The phagocytosis assay was performed using the Vibrant Phagocytosis assay kit following the manufacturer's protocol. RAW264.7 cells were used as control. Briefly, HCE, IOBA-NHC, and RAW264.7 cells were grown in 96-well plates until confluence. Fluorescent *Escherichia coli* bioparticles were opsonized by incubation with human serum for 1 h at 37 °C.

The fluorescent bioparticle suspension was added to the cultures and incubated at 37 °C. Negative controls included incubation without bioparticles. After 2 h, the bioparticle suspension was removed and a Trypan blue solution (0.25 mg/ml) was used to quench the extracellular fluorescence. The fluorescence associated with the phagocytosed bioparticles inside the cells was measured in a SpectraMAX M5 multidetection microplate reader at an excitation wavelength 480 nm and an emission wavelength of 529 nm. Results were expressed as relative fluorescence units (RFU). Each assay was performed five times in triplicate.

IMMUNOFLUORESCENCE ASSAYS

HCE and IOBA-NHC cells were seeded onto eight-well multichamber Permanox slides (5×10^5 cells/well), grown until confluence, and incubated with 100 μl of 20 μg/ml HA-CSO NPs. After 1 h, the cells were washed three times with 0.27% glucose in PBS and fixed in ice-cold methanol. After several washes with PBS, they were incubated at room temperature (RT) for 50 min with blocking buffer composed of PBS with 4% goat serum, 0.3% Triton X-100, and 1% BSA to block the non-specific binding.

Then they were incubated with anti-caveolin-1 (0.25 μg/ml) or anti-clathrin antibody (10 μg/ml) at 37 °C for 1 h. After several washes with PBS, cells were incubated with PE-Cy5-conjugated secondary antibodies (4 μg/ml) for 1 h at RT. Cell nuclei were counterstained by Hoescht dye. The preparations were viewed under an epifluorescence microscope. Each experiment was performed three times, and negative controls included the omission of primary antibodies.

DEGRADATIVE PATHWAYS

Lysosomes in living cells were labeled and tracked by the fluorescent acidotrophic probe LysoSensor. HCE and IOBA-NHC cells were grown on eight-well multichamber Permanox™ slides (5×10^5 cells/well) until confluence and then incubated with HA-CSO NPs for 1 h. After incubation, the medium was removed and Lysosensor (150 nM in DMEM culture medium) was added. Cells were incubated for 3 min in 5% CO_2 at 37 °C. Before observation by microscopy, the solution was replaced with fresh DMEM medium.

TRANSFECTION ASSAYS

The yield of gene expression was evaluated by monitoring concentrations of SEAP in cells exposed to pSEAP-loaded HA-CSO NPs, using the fluorimetric Sensolyte 3,6-fluorescein diphosphate (FDP) SEAP reporter gene assay. Samples of culture medium were prepared according to the manufacturer protocols 48 h after NP incubation. The fluorogenic phosphatase substrate FDP was used to assess the activity of generic phosphatase activity.

STATISTICAL ANALYSIS

Statistical analyses were performed by a biostatistician (co-author I.F.). Results were expressed as means±standard error of the means. The statistical significance between control NP treatment and inhibitor effects at each time point was analysed by ANOVA with three fixed effects, using the Box-Cox transformation with $\lambda=-1$ to verify the assumptions of normality, variance homogeneity, and independence. Data from phagocytosis, transfection, and viability assays were analysed by ANOVA with two fixed effects. After performing a Levene's test to assess the equality of variances and a variance decomposition to determine how much of the forecast error variance could be explained by the model, pairwise comparisons were performed. Differences were considered to be significant when pd≤0.05.

DEVELOPMENT OF VALID/REALISTIC TOXICITY TESTING PROTOCOLS

No single method is expected to substitute for the complexity of systemic or local toxicity in humans or animals. The response of the mammalian organism to a chemical is known to involve various physiologic targets and a variety of complex toxicokinetic factors. In addition, humans interact simultaneously with a variety of toxic insults, logarithmically increasing the toxicologic outcomes *in vivo*. Thus, many questions have emerged as a result of historical and current attempts at developing *in vitro* and alternative models for the prediction of systemic and local toxicity.

For instance, why should an *in vitro* test be expected to respond to the same number of toxic insults as the animal model, particularly when an animal test is not called upon to explain the myriad effects from the combination of several chemicals? Thus the answers to these queries are not limited by the biotechnology available now for the development of alternative models, but by the lack of a cohesive effort to implement the objectives of the original "3R" plan (reduction, refinement, replacement).

That is:

- Are alternative methods capable of fulfilling the need to model different types of quantitative general toxicity, such as acute systemic toxicity or local irritancy?
- Can the information obtained from *in vitro* models be used as a paradigm for predicting, supplementing, replacing or confirming animal toxicology data?

  These two questions may require an assortment of clever approaches. For instance, it is possible to apply the 3Rs to animal toxicology testing[1] of existing chemicals with the results generated from *in vitro* tier testing. This goal is achievable with an amalgamation of results obtained from reliable chemical and analytical measurements of different samples—for example, information obtained from human and animal blood and tissue concentrations of the same chemicals, from human volunteer testing of dermal and ocular irritancy, or through the tabulation of databases with clinically relevant human toxicity information. Based on observable trends and practical experience of many laboratories and regulatory agencies with *in vitro* cytotoxicology programmes, it is conceivable and realistic that toxicology testing in the foreseeable future will be regularly performed with cell culture models. Furthermore, such testing will be more efficient and more predictive of human toxicity than the current animal tests, given the amount of time, resources, and variability of information associated with animal testing. With the realization of this objective comes the benefit that many new types of *in vitro* toxicology and toxicokinetic tests will be available, while many of the current methods will be validated in refined programmes. These programmes are conducted as refined multi-factorial model systems of large *in vitro* databases to account for human toxicity, including computerized physiologically based kinetic modeling. Simultaneously, the gradual acceptance of new *in vitro* methods should not only depend on formal validation programmes, but may be a consequence of the parallel experience of various academic, government, regulatory and industrial efforts using both *in vitro* and *in vivo* methods. Finally, these achievements are based not only on political or societal attempts to reduce, refine, or replace the use of animals in research, but on the complementary approaches of efficient and reliable systems already in place for protecting the public interest.

- Can the results generated from these methods be used to compile a set of standardized tests which together can substitute as the predictive battery?

  This goal may require several relevant systems, such as human hepatocytes, heart, kidney, lung, nerve cells, and other cell lines of applicable and relevant importance. Twenty years ago, the fledgling perception that a standardized battery of *in vitro* tests could be developed for tier testing of acute local and systemic toxicity, was ambitious. Progress in sensitivity, reliability, and technical feasibility of testing protocols, supported by current validation studies, are a testament to the realization of those original objectives. Certain methods assigned to the battery are used for systematic mapping of the different effects attributed to general systemic cytotoxicity. The mechanisms underlying cytotoxic phenomena explain how these effects are influenced by the physicochemical properties of the molecules and how the knowledge is useful for the interpretation of results in cellular testing. Alternative cell based models are applied on actions of chemicals known to be cytotoxic to animals or humans by comparing the *in vitro* and *in vivo* concentrations of the same chemicals.

  Figure 7.1 summarizes the organization and applications of *in vitro* cytotoxicity testing and assigns a position for systemic cytotoxicity, organ-specific cytotoxicity, and organizational cytotoxicity within the general scheme. According to the diagram, carcinogenicity and mutagenicity data contributes to cytotoxicology evaluation but this information can be circumvented. Note also that *in vitro*toxicokinetic studies contribute to the establishment of the model. Together with *in vitro*concentrations derived from the tests, a model for cytotoxicity is formulated. Overall, the objective is to arrive at human toxic blood concentrations to assist in the prediction of human toxicity and evaluation of risk assessment.

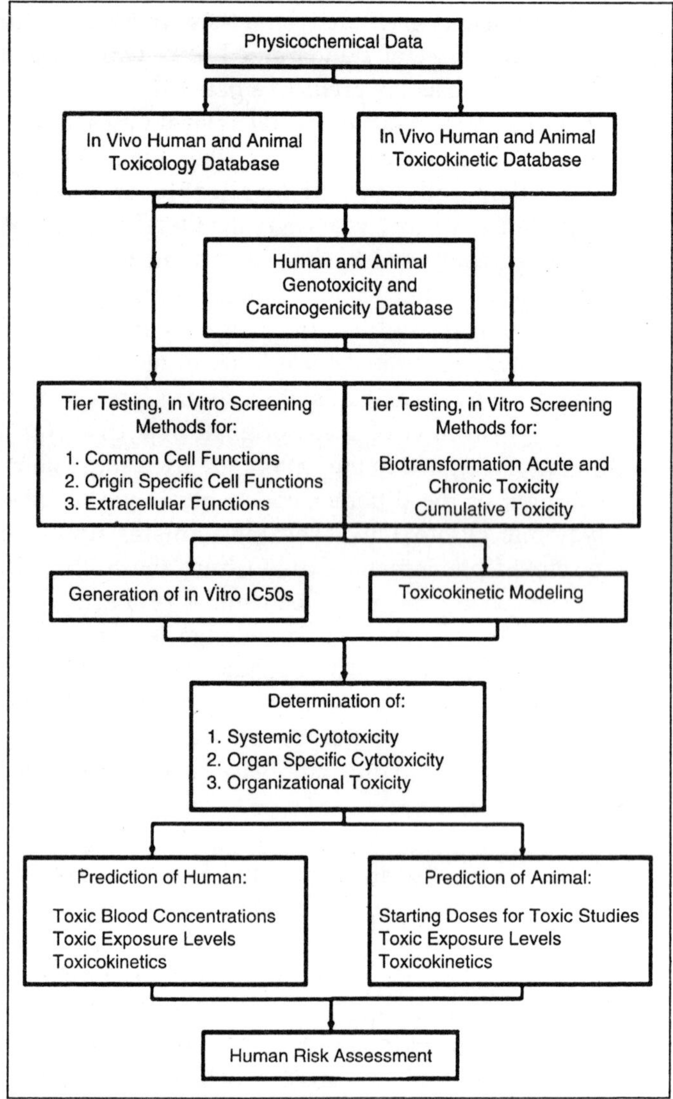

Fig. 7.1 Summary of the Organization and Applications of *in Vitro* Cytotoxicity Testing with Assignment of Positions for Systemic, Organ-specific, and Organizational Cytotoxicology within the Scheme of Risk Assessment.

Accordingly, a battery of selected protocols functions as a primary screening tool before the implementation of routine animal testing. Subsequent animal tests are then performed to confirm the results from the tier testing, as well as to detect outliers that elude the screening protocols. Upon completion of a specific and limited number of animal tests to confirm the *in vitro* data, a summary report of all available information is used as a basis for screening and/or predicting risk to humans.

With the refinement of general *in vitro/in vivo* testing and the progress afforded such studies with time, certain tests will be employed with greater confidence. This is forecasted as it becomes apparent that the battery reflects the acute or chronic toxicity of certain groups of substances, resulting in the inevitable elimination of particular aspects of animal testing.

4. Are quick, simple, and economic *in vitro* cell systems capable of sustaining the regulatory standards demanded by public health needs?

   Together with genotoxic and mutagenic tests, *in vitro* cellular test batteries are applied as biological markers of risks induced by chemicals, whether synthetic or naturally occurring. Such risks include biomonitoring of food and food additives and water and air analysis for environmental toxicants. If similar batteries of protocols are validated for their ability to screen for or predict human toxic effects, these measurements will establish a significant frame of reference for monitoring of environmental, occupational and commercial toxic threats.

   Although much of the mystery originally surrounding the techniques has waned, the doctrine underlying the establishment of the methodology remain. For instance, improvement in serum requirements in media has fostered continued excitement in the role of growth factors and cell differentiation. In addition, had it not been for these techniques, the recent discoveries in stem cell biology may not have been realised. Consequently, the discipline of cell culture, and its application to *in vitro* toxicology and biology, represents a tool poised to answer scientific questions in the biomedical sciences that can only be addressed with the proliferation of isolated cells, without the influence of other organ systems. With this understanding, the use of cell culture does not purport to represent the whole human organism, but can significantly contribute to our understanding of the workings of its components.

- Is the collaborative effort primed to secure a significant impact on the 3Rs?

  Historically, it was discovered that a tissue or organ that was part of an intact biological specimen, when isolated, would exhibit a breakdown in the supporting matrix, followed by migration of individual cells from the specimen as a consequence. With the addition of serum or whole blood, the resulting clot formed a hanging drop, making it possible to look at these cells through an ordinary light microscope. With some refinement, the "explants" demonstrated that cells migrate out of the dissected specimen. Common knowledge dictates that organs are composed of tissues, and cells are their framework. This decades-old axiom was re-discovered when it became possible to cultivate individual cells in small glass tubes while bathed in plasma or embryonic serum extract. Advances in aseptic techniques allowed cells to be maintained in culture for longer periods of time in the absence of microbiological contamination. Further developments in synthetic media, sterile supplies and equipment, and understanding of extracellular interactions were driven by the recognition that this technology was not only scientifically necessary, but that the contributions to public health would be unmistakably promoted.

Today, the same idealistic motivation presides over the toxicologic scientific goal as it did years ago—that is, can one or a few *in vitro* tests mimic the *in vivo* response.

The test(s) would require the following:

- Sensitive indicators capable of discriminating among results to avoid false positives and contribute an explanation towards the target mechanism (sensitivity testing);
- The cell system would need to integrate the essence of visceral organs to express the myriad of possible responses (testing for relevance);
- The length of time of *in vitro* exposure would be equivalent to *in vivo* time periods (testing for reliability); and,
- The model system should generate toxicokinetic and toxicodynamic interactions reproducible in many laboratories (reliability).

To date, many proposed alternative *in vitro* toxicology tests in the U.S. have accumulated in the scientific literature and regulatory platforms without the benefit of arriving at general acceptance. The situation is due in part to the difficulty associated with introduction of experimental systems from individual laboratories. Thus the organization of multilaboratory validation programmes is prompted by *in vitro* toxicologists as a possible remedy for the problem. For example, public interest groups, academic institutions, corporate industries, and government agencies in the European Union (E.U.) and U.S. must be actively engaged in promoting the development of *in vitro* toxicology tests by organizing well planned programmes.

Contemporary validation studies can be facilitated through collaborative international cooperation, particularly through the efforts of government sponsored and independent organizations, such as the U.S. Interagency Coordinating

Committee on the Validation of Alternative Methods (ICCVAM), U.S. National Toxicology Programme (NTP) Interagency Center for the Evaluation of Alternative Toxicological Methods (NICEATM), the European Centre for Validation of Alternative Methods (ECVAM), and the Center for Documentation and Evaluation of Alternative Methods to Animal Experiments (ZEBET, Germany).

These alliances provide a foundation for validating alternative methods along with promotion and encouragement for the harmonization of scientific approaches to validation and review. In addition, both ICCVAM and ECVAM have outlined key objectives: to stimulate development of test methods and testing strategies in prioritized areas by individual laboratories and to facilitate the nomination of promising test methods for appropriate regulatory translation.

CONCLUDING REMARKS

Current and future validation should not be hampered by theoretical considerations of the best way to achieve regulatory acceptance. The refinement of the validation programmes should be approached as a set of generally applicable rules. The most effective future validation will probably require studies directed at all levels, including the efforts of individual laboratories, reliability-oriented multilaboratory programmes, and multiple-method multilaboratory perspectives focused on cell based toxicology. In addition, methods used extensively in past years should not be treated with the same approach as newly developed, untried procedures. This perspective wastes time and resources.

It is important to understand that the progress of *in vitro* cytotoxicology relies on the continuous development of better strategies to evaluate methodologies rather than as a set of inflexible rules for the inclusion or exclusion of protocols. More importantly, efforts to evaluate the scientific integrity of methods for validation purposes should be managed, guided and administered by independent investigative groups rather than government or industry sponsored organizations. The groups must be free of vested interests, with their services directed towards the development of scientifically valid programmes. Consequently, the rate-limiting step is not dependent on the available biotechnology but relies on a cohesive effort among the competing entities to discover the most efficient means to this end.

8

Exposure to Nanoparticles

INTRODUCTION

Semiconductor nanoparticles, also known as quantum dots (QDs), provide extremely good fluorescent signal, and can be well used for fluorescent labeling. Also they are ideal bioconjugated fluorescent probes for bioanalysis, because of their unique optical and electronic properties, such as broad excitation spectrum, narrow emission spectrum, good tunability and photochemical stability, neglectable optical bleaching, etc. Hence, QDs have attracted considerable interest in biological and medical community. The basic knowledgement and characteristics of QDs,their synthesis and labeling approach,recent progress and prospect in potential application to immunoassay and rapid diagnostic analysis, the basic properties, synthesis, applications such as bioseparation of magnetic nanoparticles are also reviewed. A method for hydrothermal synthesis of CdTe QDs under microwave-assisted condition is developed.

Using L-cysteine hydrochloride and glutathione as stabilizing agents, the molar ratio is 2.5:1. CdTe QDs with controllable photoluminescence wavelength from 510 nm to 670 nm were prepared in 150 min, the photoluminescence quantum yield was shown to be 65%. Compared with CdTe QDs prepared with thiohydracrylic acid as stabilizing agent, as-prepared CdTe QDs show much narrower photoluminescence FWHM (Full Width at Half Maximum), more symmetrical emission peak and higher photoluminescence quantum yield. The as-prepared CdTe QDs contain–NH_2 and–COOH groups, which have good biological compatibility and have almost no toxicity. The C. elegans were taken as detecting object to study the biological toxic effect of CdTe QDs.

A novel technique of synthesizing fluorescence-encoded microspheres by dispersing polymerization method is then reported. By the following one-step swelling procedure, the different CdTe QDs were carried into the inner of polystyrene microspheres quantificationally. This procedure has the following outstanding advantaged: it can be operated easily and reproducibly. Besides, as-synthesized photoluminescent microspheres have better photostability, the CdTe QDs can not be isolated or leaked out easily which are carried into the inner of microspheres.

Finally, the multiple CdTe QDs were embedded into one silica microsphere by reverse microemulsion method. The as-synthysized silica microspheres have high luminous efficiency and small side effects on cells or living beings. A series of characterizations show that as-prepared magnetic polystyrene microspheres have narrower size distribution with smooth surface, fine morphology and structure.

The factors that affect the size of microspheres were also discussed. The as-prepared microspheres have narrower size distribution with smooth surface, fine morphology and structure, pretty superparamagnetic behaviour, the specific saturation magnetization of them reaches to 11.61 Am^2/kg at room temperature, the method can overcome these shortcomings: nonhomogeneous distribution of the interior magnetic matericals, irregular morphology of microspheres. $Fe_3O_4 \cong SiO_2$ core-shell microspheres were synthesized by reverse microemulsion method.Since Liedberg developed the first surface plasmon resonance (SPR) sensor, the research and application of SPR technology have shown extensive growth and gradually become the research hot spot in the biosensor field in the world. The most important characteristics of SPR sensors are their versatility and capability for real time monitoring the association or dissociation of biomolecules on the surface of the sensor without the need for fluorescence or labeling of the biomolecules. These characteristics made the SPR technique an easy, convenient and reliable one for determining the concentration and molecular weight, monitoring change in structure, measuring kinetic constant and binding specificity of individual biomolecules.

For above reasons, the SPR sensors grow as a new powerful technology in chemical and biological field. The categories and applications of wavelength modulation SPR sensors are discussed. The recent research progress of SPR sensor is also reviewed. The setup consisted of white LED, light transmit system, sensing element, flow cell, detection system and data processing system. The light emitted from the white LED is polarized to obtain polarized light, and two lenses are employed to make the light parallel.

The parallel polychromatic light beam passes through an optical prism and excites surface plasmon at the interface between the gold film and solution. The output light from the prism is guided into a charge coupled device (CCD) detector by fibre. The sensor was shown to relatively simple to use, rapid to respond, and cost-effective.In this, the highly specific interaction of an avidin-biotin system was explored for immobilization of target DNA on the sensor memberane, $CdTe\cong SiO_2$ core-shell microspheres coupled with target DNA, the wavelength modulation SPR instrument was developed for real-time determining the immobilizing process of biotinylated DNA and the performance of hybridization reaction. The target DNA coupling with $CdTe\cong SiO_2$ core-shell microspheres was determined to be in the concentration range of 0.0262-14 nmol/ L. In addition, the method of SPR sensor regeneration was investigated (using a solution of 0.1 mol/L H_3PO_4, 0.1 mol/L NaOH, 0.1 mol/L HCl, 0.03 mol/L HNO_3 or 0.3 mol/L citrate solution). The citrate buffer (0.3 mol/L) solution was chosen as regeneration solution.

Compared with target DNA coupled with magnetic polystyrene microspheres, the silica-coated magnetic nanoparticles conjugating with target DNA was validated to have faster association rate and smaller RSD. Besides, methods of enhancing SPR sensitivity were investigated by using magnetic microspheres. The target DNA was determined solely in the concentration range of 14-100 nmol/L, while target DNA conjugating with magnetic microspheres was tested in the concentration range of 24.3-120 pmol/L.

It was shown that the association constant of biotinylated DNA probe with the target DNA coupling with magnetic microspheres was 3.31×10^7 mol-1 L, a three orders of magnitide improvement as compared with that obtained without magnetic microspheres. The 0.01 mol/L NaOH was chosen as regeneration solution. The research has greatly significance of enhancing sensitivity of SPR sensor, it reveals new direction of nanotechnology in biosensor application.

PHYSICAL CHARACTERISTICS AND PROPERTIES OF NANOPARTICLES

INTRODUCTION

Pulmonary drug delivery has become a well-established approach in the treatment of respiratory diseases and offers several advantages over other routes of administration. Inhalation therapy enables the direct application of a drug to the respiratory tract.

The "local" or "regional" deposition of the administered drug facilitates a targeted treatment of respiratory diseases avoiding high-dose exposures to the systemic circulation. With the direct delivery of therapeutic agents to the desired site of action, rapid onset of drug action, lower systemic exposure, and consequently, reduced side effects can be achieved. Site-specific or targeted delivery, therefore, would also enable a reduction in the necessary dose to be administered. A significant disadvantage of inhalation therapy is the relatively short duration of drug action demanding multiple daily inhalation maneuvers, ranging up to 9 times a day. Moreover, "conventional" inhalation therapy does not permit targeted cell-specific drug delivery or modified biological distribution of drugs, both at the organ and cellular level, and drug deposition in different lung areas is only poorly controllable.

Strategies for further advancements of inhalation therapy include the development of aerosolizable controlled release formulations with the aim to improve the drug effect, as well as the patient's convenience and compliance. A large number of carrier systems have been conceived and investigated as potential controlled drug delivery formulations to the lung.

In the recent years, nanomedicine has become an attractive concept for the controlled and targeted delivery of therapeutic and diagnostic compounds to the desired site of action. Nanotechnology opens new perspectives in the design of novel drug delivery vehicles that not only facilitate targeting of an organ, tissues, cells, or subcellular compartments, but also affect the duration and the intensity of the pharmacological effect. In particular, nanoparticulate drug delivery systems enable the controlled delivery of the pharmacological agent to its site of action at a therapeutically optimal rate and dose regimen.

Among the various drug delivery systems considered for pulmonary application, polymeric nanoparticles demonstrate several advantages for the treatment of respiratory diseases, for example prolonged drug release and cell-specific targeted drug delivery. Numerous manufacturing techniques are known for the production of drug-loaded polymeric nanoparticles. The choice of the nanoparticle preparation technique essentially depends on the physicochemical properties of the polymeric nanoparticle matrix material intended to be used and on the active compound to be encapsulated in the nanoparticles. Regarding the polymeric nanoparticle matrix material, criteria such as biocompatibility and degradability determine its selection.

Moreover, for an effective nanoparticulate drug delivery system, sufficient drug loading and controlled drug release over a predetermined period of time must be ensured. The characteristics of drug release, that is, release mechanism and release rate, from drug-delivery systems vary according to the type of employed encapsulation technique and the physicochemical properties (interaction) of drug and polymer. The release from polymeric nanoparticles *in vitro* is normally fast (several minutes to hours) due to the short distance drugs have to cover to diffuse out of the particles.

The release rate of drugs from nanoparticles is also strongly influenced by the biological environment. Nanoparticles may interact with biological components like proteins and cells that alter the release rate of drugs from nanoparticles. As a consequence, the *in vitro* drug release characteristics may not predict the release situation *in vivo*. Moreover, a precise assessment of the *in vitro* drug release from nanoparticles is technically difficult to achieve, which is mainly attributed to the inability of rapid separation of the nanoparticles from the dissolved or released drug in the surrounding medium.

Different methods have been used to characterize the behaviour of pulmonary administered drug-loaded carriers in biological systems. These range from *in vitro* cell culture methods to *in vivo* pharmacokinetic analysis. *Ex vivo* isolated, perfused, and ventilated lung models have been utilized in numerous pharmacological and toxicological studies to elucidate the fate of inhaled drugs or toxic substances.

In*ex vivo* lung models, lung-specific pharmacokinetic effects, like drug absorption and distribution profiles, can be investigated without the contribution of systemic absorption, distribution, and elimination of the drug. Moreover, it is possible to elucidate the effect of the interaction of nanoparticles with the natural pulmonary environment on the release of encapsulated drugs. Accordingly, more reliable drug release and distribution data are obtained that are closer to the *in vivo* situation.

It is interesting to note that the first investigations regarding the use of polymeric nanoparticles as drug carriers for the controlled and targeted delivery of drugs to the desired site of action have been reported in the mid 1970's. It was shown that the "natural" drug distribution after systemic application was altered by the encapsulation of drug into polymeric nanoparticles. Since then, great efforts have been made in this field, and several treatment modalities for cancer on the basis of polymeric nanostructured drug delivery vehicles have been developed and made clinically available, for example, Abraxane, Transdrug, and Genexol-PM. In contrast to systemic administration, the regional application of drug-loaded nanoparticles to the respiratory tract has been so far incompletely investigated.

This is attributed on one hand to the limited efficiency of conventional devices to generate nanoparticle-containing aerosols and on the other hand to the lack of methods to assess the drug release form and the distribution behaviour of pulmonary administered nanoparticulate drug-delivery systems. Meanwhile, technological advances have led to improved designs for aerosol-generation devices that solve the main drawbacks, and the key attributes associated with successful nanoparticle aerosolization have been identified. However, the prediction of drug release and distribution from pulmonary administered nanoparticulate-delivery systems remains a major challenge.

STRUCTURE AND FUNCTION OF THE LUNG

The development of drug delivery systems for pulmonary application requires a detailed knowledge of the lung in its healthy, as well as various diseased states. The lung is composed of more than 40 different cell types, of which approximately one-third are epithelial cells. The conducting zone includes the nasal cavity, pharynx, larynx, trachea, bronchi, and bronchioles, while the respiratory zone, where the gas exchange takes place, includes respiratory bronchioles and alveoli. The conducting airways exhibit 16 bifurcations, comprising the trachea, the bronchi, and the bronchioles. The terminal bronchioles represent the passage to the respiratory region, which exhibits another 6 bifurcations.

The respiratory region includes the respiratory bronchioles, from which the alveolar ducts with alveolar sacs branch off. The airways also fulfill some other essential functions, such as warming, humidifying, and cleaning of the inhaled air. Warming and humidifying of the inspired air predominantly take place in the nasal cavity and the pharynx. In the deeper airways this process continues, so that the air finally reaching the alveoli has body heat and is completely saturated with water. Also the cleaning of the inspired air partly takes place in the nose; dust, bacteria, and particles are caught by impaction.

Further inhaled substances deposit on the mucus layer which coats the walls of the conducting airways. The mucus is secreted by goblet and submucosal gland cells and forms a gel like layer consisting of mucin as the major component. Ciliated cells are another important type of cells which predominate in the bronchial epithelia of the conducting region. Their major function is the propulsion of mucus upwards and out of the lung (bronchotracheal escalator), thus the lung will be cleared of foreign substances. Beneath, in the respiratory bronchioles the epithelium consists of ciliated cells and Clara cells.

In the alveolar space there is no mucus layer, but a complex surfactant lining that covers the alveolar epithelium and reduces the surface tension to prevent collapse of the alveoli during breathing. It contains approximately 90% lipids and 10% proteins. The lipids in the surface lining material consist mainly of phospholipids (~80–90%) and a minor portion of neutral lipids (~10–20%). Among the phospholipids, phosphatidylcholines (~70–80%) and phosphatidylglycerols (~10%) represent the predominant classes, with minor amounts of phosphatidylinositols, phosphatidylserines, and phosphatidylethanolamines. About half of the protein mass of the alveolar lining layer is composed of the surfactant-associated proteins SP-A and SP-D, which are high molecular weight hydrophilic proteins, and SP-B and SP-C, which are low molecular weight hydrophobic proteins.

The surfactant proteins SP-A and SP-D have been identified as playing a fundamental role in innate immunity. A complex interaction between phospholipids and SP-B and SP-C enables the decrease of surface tension in the alveolar region to values of ~0mN/m during compression/expansion cycles. Pulmonary surfactant is secreted by type II pneumocytes, which cover only 5% of the total alveolar surface. Beside the production of pulmonary surfactant, alveolar type II cells play a role in alveolar fluid balance, coagulation and fibrinoylsis, host defence and proliferation, and differentiation into type I cells. Type I pneumocytes are very thin (≤200nm) with a large extension (~200μm), covering over ~95% of the alveolar epithelial surface. They form the primary diffusion barrier between air and blood which is highly permeable for water, gases, and hydrophobic molecules, while it is poorly permeable for large hydrophilic substances (peptides and proteins) or ionic species.

Macromolecules pass this barrier by active transport mechanisms. In addition to epithelial cells, the alveoli contain macrophages that engulf particles, potentially digest them, and slowly migrate with their payload out of the respiratory tract, either following along the mucociliary escalator or (to a lesser degree) the lymphatic system. Thus, the pulmonary endocytosis by macrophages represents the main mechanism of removing solid particles in the alveolar region. Nanoparticles have been praised for their advantageous drug delivery properties to the lung, such as avoidance of mucociliary and macrophage clearance and long residence times until degradation or translocation by epithelial cells takes place.

NANOPARTICLE EXPOSURE

Although the lung represents effective barrier systems and clearance mechanisms much attention has been raised in the last decades to this organ for drug delivery applications. One reason is its large absorption area. The lung build up a total surface of ~100m^2 that is enveloped by an equally large capillary network, from which many agents can be readily absorbed to the bloodstream avoiding a first-pass-effect of the liver.

Another reason is the known instability and low permeability of proteins and peptides when these biopharmaceuticals are administered through the widely preferred oral route. Consequently, most proteins and peptides on the market are administered intravenously. But the parenteral route of application does generally not meet with patients' convenience and compliance, in particular because the indication for the use of these agents is usually treatment of a chronic disease requiring frequent injections. Thus, the pulmonary route of application offers a noninvasive alternative for systemic therapy. However, systemic macromolecule delivery via the lung has suffered setbacks as for example demonstrated for pulmonary administered insulin (Exubera) that was withdrawn from the market in 2008 for commercial and health-risk reasons.

A large number of small molecular weight drugs are employed for the targeted treatment of respiratory diseases following inhalation. This basic concept of targeted drug therapy has been followed for a long time in the treatment of airway diseases. In particular, the application of β_2-agonists and corticosteroids by means of inhalation has improved the therapy of bronchial asthma and chronic obstructive pulmonary disease targeting the smooth musculature of the bronchi and immunologically competent intrapulmonary cells. In addition, the endothelial cells or the smooth muscle cells surrounding the pulmonary vessels present a target of inhalative drug therapy. As an example, prostaglandin derivatives have been recently introduced for aerosol therapy of pulmonary arterial hypertension.

DEVICES FOR AEROSOL GENERATION

Over the past decades several devices have been conceived and developed for the administration of drugs to the respiratory tract, namely pressurized metered dose inhalers (pMDIs), dry powder inhalers (DPIs), and nebulizers. pMDIs are hand-held devices that use pressurized propellants to atomize the drug solution, suspension, or emulsion. These devices generally require a coordinative inhalation by the patient. DPIs do not only differ in the principle of aerosol particle generation and delivery, but also with regard to design differences such as discrete or reservoir drug containment and the number of doses. While the drugs are released from pMDI by the utilization of propellants, DPIs operate by using the inspiratory flow of the patient for disintegration of the powder and dose entrainment. Thus, reproducibility of the inhaled dose from these devices is extremely dependant on the patient.

Several types of nebulizers are available for aerosol generation for pulmonary drug delivery, namely jet nebulizers, ultrasonic nebulizers, and nebulizers that use a vibrating-mesh technology for aerosol generation. Jet nebulizers are driven by compressed air. The liquid is dispersed into small droplets (<5-6μm) by passing through a narrow nozzle orifice and multiple impactions on a baffle structure. In general, the droplet size distribution of a nebulizer and the output rate are also influenced by the physical properties of the drug solution and the air flow rate from the compressor.

Ultrasonic nebulizers use a piezoelectric transducer in order to create droplets from an open liquid reservoir. As the energy is transferred through the liquid container it becomes evident that the properties of the drug formulation have strong effects on the aerosol particle size and the output rate. Vibrating-mesh nebulizers use perforated membranes actuated by an annular piezoelement to vibrate in resonant bending mode. The holes in the membrane have a large cross-section size on the liquid supply side and a narrow cross-section size on the side from where the droplets emerge. Depending on the therapeutic application, the hole sizes (2μm and upwards) can be adjusted, as well as the number of holes.

POLYMERIC NANOPARTICLES AS INHALATIVE DRUG-DELIVERY VEHICLES

Nanomaterials exploit novel physical, chemical, and biological properties. The general aim of controlled release formulations is the modification of pharmacokinetics and thus, improved pharmacodynamic characteristics at the target site. A successful drug delivery system needs to demonstrate optimal drug loading and release properties, and low toxicity. Nanoparticle formulations for this purpose with a mean size between 50 and 300nm normally consist of polymeric materials.

Polymers with particular physical or chemical characteristics, such as biocompatibility, degradability, or responsiveness to environmental changes have been predominantly used. In addition to biocompatibility and degradability of the applied polymer, sufficient association of the therapeutic agent with the carrier particles and controlled and targeted drug release properties, nanoparticles need to meet further standards, such as protection of the drug against degradation, ability to be transferred into an aerosol, and stability against forces generated during aerosolization. Nanoparticles composed of biodegradable polymers fulfill the stringent requirements placed on these delivery systems.

Due to their well-established biocompatibility and biodegradability, aliphatic polyesters like polylactide (PLA) and poly(lactide-*co*-glycolide) (PLGA) are the most extensively used materials for biomedical applications. However, linear polyesters have many limitations as nanoparticle matrix materials. Firstly, PLGA nanoparticles degrade over a period of weeks to months, but typically deliver drugs for a much shorter period of time.

Slow or nondegrading polymers may lead to an unwanted accumulation in the lung when repeated administrations are needed, and may cause inflammatory processes. One way to overcome this problem is to synthesize polymers with faster degradation rates. Fast-degrading polymers are obtained by grafting of short PLGA chains onto polyvinyl alcohol backbones. The adjustable properties of these branched polyesters make them highly suitable for pulmonary formulations, especially with regard to biodegradation rates and *in vitro* cytotoxicity.

Moreover, these types of biodegradable polyester revealed no signs of inflammatory response *in vivo*. Their amphiphilic properties allows the generation of nanoparticles without the use of additional surfactant stabilizers. Another type of biodegradable polymer suitable for pulmonary application is based on ether-anhydride terpolymers consisting of poly(ethylene glycol), sebacic acid, and 1,3-bis(carboxyphenoxy) propane. These polymers are known to form aerosolizable particles and to exhibit fast degradation rates (half-life <12h). Secondly, for an effective nanoparticulate-delivery system, sufficient drug-loading and tailored release properties must be ensured. Nanoparticles prepared from hydrophobic polymers, like PLGA, often incur the drawback of poor incorporation of low molecular weight hydrophilic drugs due to the low affinity of the drug compounds to the polymers. The introduction of charged functional groups within the polymer structure, like for example described by Wittmar *et al.* and Wang *et al.*, promotes electrostatic interactions with oppositely charged drugs, thereby improving the design of nanoparticulate carriers.

The release rate and release mechanism from drug-delivery systems vary according to the carrier vehicle, as well as to the properties of the employed drug and polymer combination. The *in vitro* release pattern from polymeric nanoparticles used in the field of medicine and pharmacy is of importance for characterization purposes and for quality control reasons. The release of drug compounds from nanoparticulate drug delivery systems is a result of the direct interaction of nanoparticles with their environment and is thought to be dependent upon desorption of the surfacebound, adsorbed drug, diffusion through the nanoparticle matrix, and rate of polymer degradation. Thus, diffusion and biodegradation govern the process of drug release from polymeric nanoparticles.

Several manufacturing techniques are known for the production of drug-loaded polymeric nanoparticles, allowing extensive modulation of their characteristics and control of their behaviour at the target site. Conventionally, two groups of preparation methods can be distinguished.

The first involves polymerization of monomers whereas the second is based on precipitation of preformed, well-defined natural or synthetic polymers, as for example used in salting out, emulsion evaporation, emulsification diffusion, and solvent displacement. The choice of the nanoparticle preparation technique essentially depends on the physicochemical properties of the polymeric nanoparticle matrix material intended to be used and on the active compound to be encapsulated in the nanoparticles. One way to encapsulate the drug into the nanoparticles is accomplished by the preparation of nanoparticles in the presence of the therapeutic agent, what leads to a "homogeneous" distribution of drug within the polymer matrix. Another way to associate drug and polymer is achieved by subsequent sorption of the drug to unloaded nanoparticles either to the surface or the bulk of nanoparticles. The type of binding may also result in different release mechanisms and release rates.

Overall, the final choice of the appropriate polymer, manufacturing technique, and nanoparticle characteristics will primarily depend on the biocompatibility and degradability of the polymer, secondarily on the physicochemical characteristics of the drug, and thirdly on the therapeutic goal to be reached.

Owing to the advantageous drug delivery properties of polymeric nanoparticles, researchers were encouraged to find suitable application forms for pulmonary delivery. Their small size limits pulmonary deposition as nanoparticles alone are expected to be exhaled after inhalation. In general, aerosol particle size is characterized by the mass median aerodynamic diameter (MMAD). The MMAD is used to describe the particle size distribution of any aerosol statistically based on the weight and size of the particles.

Thus, a group of very dense particles will exhibit a larger MMAD than that of a group of less dense particles, despite an identical geometric size. It is well understood that pulmonary deposition is achieved by three principal mechanisms: inertial impaction, sedimentation, and diffusion. Impaction predominates during the passage through the oropharynx and large conducting airways if the particles possess a MMAD of >5μm, or have a high velocity. Gravitational force leads to sedimentation of smaller particles (MMAD of <3μm) in the smaller airways.

Additionally, sedimentation increases by breath holding. In the range below a MMAD of 1μm, particles are deposited by diffusion, which is based on Brownian motion. Thus, extent and efficiency of drug deposition is influenced by particle-specific and physiological factors, such as particle size and geometry, lung morphology, and breathing pattern. Common methods to deposit drug-loaded nanoparticles in the deeper lung are the nebulization of nanosuspensions and the aerosolization of nanoparticle-containing microparticles (composite microparticles).

A number of nanoparticle formulations were found to be accessible for nebulization with common nebulizers. One major advantage of this method is that regardless of the aerodynamic properties of the nanoparticles themselves, alveolar deposition can be easily achieved by generating adequate droplet sizes. Over the past decades, the generation of therapeutic aerosols has primarily been reserved to pneumatic- and ultrasound-driven nebulizers. Recent technological advances have led to improved nebulizer designs employing vibrating-mesh technology for aerosol generation. Vibrating mesh nebulizers have been shown to overcome the main drawbacks of pneumatic- and ultrasound-driven nebulizers, that is, concentration of medicaments, temperature changes, and high residual volumes inside the nebulizer reservoir. The aggregation of nanoparticles during aerosolization is dependent on both the nanoparticle surface characteristics and the technique for aerosol generation.

The aggregation tendency was reduced for nanoparticles exhibiting a more hydrophilic surface. Coating of nanoparticle surfaces with hydrophilic polymers was also shown to improve the nebulization stability of biodegradable nanoparticles. Furthermore, the use of vibrating mesh nebulizers is suitable for the delivery of "delicate" structures, like biodegradable nanoparticles due to avoidance of high shear stress during aerosolization.

As an alternative to nebulization of a nanosuspension, polymeric nanoparticles can be delivered to the lung by means of dry powder aerosolization. For this reason, nanoparticles need to be encapsulated into composite microparticles using standard techniques like spray drying or agglomeration. The composite microparticles must display defined aerodynamic properties (MMAD) to obtain peripheral lung deposition of inhaled particles. The delivery of nanoparticles as part of microparticles has been intensively investigated for several reasons. A common obstacle that limits the use of biodegradable polymeric nanoparticles is their chemical and physical instability in aqueous suspension.

Nanoparticles tend to aggregate when stored over an extended period of time. Furthermore, hydrolytic degradation of the polymeric nanoparticle matrix material and drug leakage from nanoparticles into the aqueous medium take place. Thus, for stabilization of biodegradable polymeric nanoparticles a subsequent drying step needs to be carried out to remove water from these systems.

The most commonly used methods to convert a colloidal suspension into solid powders of sufficient stability are freeze- and spray drying. Spray drying offers the advantage over freeze-drying that nanoparticles are transformed to respirable microparticle-containing powders in a one-step process. Freeze-drying would cause additional disintegration to form microparticles suitable for pulmonary application.

The addition of stabilizers like sugars or polymers has shown to prevent unwanted nanoparticle aggregation during drying and storage. Spontaneous redispersion of nanoparticles is a key desideratum in the development of successful composite drug delivery systems to the lung. Composite microparticles should release their therapeutic payload (drug-loaded nanoparticles) when they get into contact with aqueous media, and the unaffected nanoparticles can carry out their therapeutic benefit at the target site. Typical examples for preparation of composite microparticles by spray drying can be found in the literature. Drug-loaded PLGA nanoparticles and trehalose as microparticle matrix material were used to prepare composite microparticles suitable for inhalation.

The use of "porous nanoparticle-aggregate particles" (PNAPs) as dry powder delivery vehicles to the lung was investigated by Sung *et al.*. Drug-loaded PLGA nanoparticles were prepared using a solvent evaporation method and subsequently converted to PNAPs using a spray drying technique. Effervescent powder formulations containing nanoparticles were recently introduced for pulmonary drug delivery. These formulations were composed of poly(butyl cyanoacrylate) nanoparticles and as effervescent components sodium carbonate and citric acid stabilized with ammonia were employed. The active release mechanism (effervescent reaction) of the composite microparticles was observed when the carrier particles were exposed to humidity and unaffected nanoparticles were released.

Another interesting method to obtain nanoparticle containing microparticles is enabled by controlled agglomeration of oppositely charged nanoparticle populations. Positively- and negatively charged biodegradable nanoparticles are brought into contact under vigorous stirring, and spontaneous composite microparticle formation takes place. Nanoparticle aggregation is driven by electrostatic attraction/ forces in this case.

CONTROL OF EXPOSURE

A basic concern in the field of nanomedicine is the development of successful nanoparticulate controlled release formulations with the aim to improve the characteristics of the therapeutic agent at the target site. Biological environments are known to strongly influence the release properties of nano-sized drug delivery vehicles. An insistent problem in the development of nanoparticulate drug delivery systems is the lack of systems to follow the drug release after contact with an external medium. As a consequence, conventional *in vitro* drug release studies may have very little in common with the delivery and release situation *in vivo* and the development of more sophisticated controlled drug release carriers to the lung is precluded.

Different preclinical models are used to account for the drug release mechanisms, as well as the rate and extent of drug absorption after pulmonary administration. The complexity of the employed models increases from *in vitro* cell culture methods and *in silico* models, which are primarily used as screening tools, to *in vivo* pharmacokinetic analysis that provide fundamental information about the fate of the released drug by monitoring drug levels in plasma, lung fluid, and tissue. Several cell culture models of the respiratory tract are described using both continuous and primary cells to explore drug transport mechanisms under precise experimental conditions. Continuous cell cultures using alveolar or bronchial epithelial cells like A549 and Calu-3 are often employed as simple *in vitro*models for pulmonary drug delivery studies. In contrast to continuous cell cultures primary cultures consisting of alveolar epithelial cells present cell morphologies and biochemical characteristics closer to the *in vivo* situation. However, time-consuming isolation and cultivating, as well as limited cell lifetimes are the main drawbacks of primary cell culture models.

The estimation of drug absorption from the respiratory tract based on physicochemical properties and permeability of drugs using computational and experimental models led to the extension of the biopharmaceutical classification system (BCS). The pulmonary BCS (pBCS) takes into consideration the specific biology of the respiratory tract, particle deposition, and the subsequent process of drug absorption and depicts an alternative to the currently and widely used studies in animals. Drug structure-permeability relationships may contribute to reliable prediction of pulmonary pharmacokinetics used for the development of novel inhalable drugs. *Ex vivo* isolated, perfused, and ventilated lung models allow the investigation of lung-specific pharmacokinetic effects on the fate of inhaled therapeutics.

These preparations maintain structural integrity of the lung tissue and allow careful control of the experimental regimen of the isolated lung. Drugs can be administered directly to the respiratory tract in a quantitative and reproducible manner, and simple sampling and analysis of perfusate provide the absorptive profile. The fundamental information about the fate of the inhaled therapeutics gained from *in vivo* pharmacokinetic analysis are accompanied by reduced screening capacity, increased expense, ethical considerations, and the potential for nonlinear dose-response relationships between the *in vitro* and *in vivo* situation as described for inhaled toxic substances. Overall, the application and comparison of different models to elucidate the drug behaviour at the target site is needed to establish reliable *in vitro-in vivo*correlations.

The preparation of the IPL using a rabbit as organ donor animal is described briefly hereafter and interested readers are referred to excellent reviews on this topic. In the following, the method of Seeger *et al.* is briefly described. For lung isolation animals need to be deeply anesthetized and anticoagulated. Then a median incision is made to expose the trachea by blunt dissection, and a cannula is inserted into the trachea. Subsequently, the animals are ventilated with room air, using a respirator.

After mid-sternal thoracotomy, the ribs are spread, the right ventricle is incised, and a fluid-filled perfusion catheter is immediately placed into the pulmonary artery. Immediately after insertion of the catheter, perfusion with cold buffer fluid is started, and the heart is then cut open at the apex. Next, the trachea, lungs, and heart are excised *en bloc* from the thoracic cage. A second perfusion catheter with a bent cannula is introduced via the left ventricle into the left atrium and is fixed in this position. After rinsing the lungs with buffer fluid for washout of blood, the perfusion circuit is closed for recirculation. Meanwhile, the flow is slowly increased, and left atrial pressure is set to 1.5mmHg.

In parallel with the onset of artificial perfusion, ventilation is changed (5% CO_2, 16% O_2, and 79% N_2) to maintain the pH of the recirculating buffer at 7.4. Tidal volume is 10ml/kg body weight with a frequency of 30strokes/min. The IPL is placed in a temperature-equilibrated housing chamber (37°C), freely suspended from a force transducer for continuous monitoring of organ weight. Pressures in the pulmonary artery, the left atrium, and the trachea are registered by means of small-diameter tubing threaded into the perfusion catheters and the trachea and connected to pressure transducers. Only lungs that have a homogeneous white appearance with no signs of hemostasis, edema, or atelectasis, a constant mean pulmonary artery and peak ventilation pressure in the normal range (4–10 and 5–8mmHg, resp.), and are isogravimetric during an initial steady state period of at least 30min are considered for experiments.

A number of visual and physiological parameters can be determined to ascertain the viability of the IPL preparation. Under optimized preparation and perfusion conditions, greater than 85% of all excised lungs fulfill these criteria, and lungs may be perfused for ~6h without changes in physiological aspects.

For analysis of pulmonary pharmacokinetics of inhaled therapeutics, such as pulmonary absorption and distribution characteristics, formulations need to be delivered to the IPL by the intratracheal route. Intratracheal delivery can be carried out by dry powder insufflation or inhalation or by means of fluid instillation or nebulization. For nebulization purpose, a nebulizer unit is connected to the inspiratory tubing between the ventilator and the lung to pass the produced aerosol through by the inspiration gas. In order to determine the absorption of drug from the lung into the perfusate, samples are taken from the venous part of the system. Additionally, the analysis of the drug distribution characteristics to the different compartments of the lung is performed by lavage for the amount of drug remaining in the lung-lining fluid and by extraction or microscopic techniques for the amount of drug remaining in the lung tissue at the end of the experiment.

APPLICATION OF IPL

The choice of an appropriate IPL preparation set-up for the analysis of the fate of inhaled therapeutics at the target site is influenced by several factors. The particular problem determines the selection of experimental parameters like ventilation method, perfusion characteristic, and perfusate type. The ventilation of the IPL can be realised by two modes, namely, "positive" pressure and "negative" (subatmospheric) pressure ventilation. During "positive" pressure ventilation a connection of a respirator directly to the lung, which pushes bolus volumes of air into the lung, is required. Subatmospheric pressure ventilation is accomplished by using a reverse connected respirator to cycle subatmospheric pressures inside the chamber in which the lung is suspended.

While "negative" pressure ventilation is generally preferred for drug absorption and distribution studies, as it prevents water loss, tissue drying, and improves organ viability (*i.e.*, reduced risk of architecture destruction by overinflation with subsequent edema formation and progressive atelectasis), "positive" pressure ventilation enables a highly efficient and homogeneous deposition of therapeutic aerosols to the isolated lung as described above. Beside the morphology of the respiratory tract, ventilation pattern is generally recognized to have a high impact for the successful aerosol delivery to isolated lungs. To avoid low dosing efficiency and nonreproducible aerosol deposition pattern in the lung, a synchronization of aerosol application and inspiration needs to be adjusted.

The applied perfusion technique is dependent on the experimental design. The perfusion may be performed in a single-pass or recirculating manner. Single-pass perfusion systems have the advantage of being less sophisticated; however, depending on the flow rate and the duration of the experiment, they come along with higher consumption of perfusate. Moreover, a sensitive sample analytics is required. In pharmacokinetic experiments of inhaled therapeutics, the use of artificial perfusion medium, for example Krebs-Henseleit buffer fluid, with addition of hydroxyethylamylopectin or dextran as oncotic agent is only appropriate for hydrophilic drug substances.

Hydrophobic drug analysis is relieved in the presence of albumin owing to binding of hydrophobic compounds. For example, Liu *et al.* investigated the effect of different perfusion buffers on the pharmacokinetics of several drugs with distinct physicochemical properties that were administered to the circulation of an isolated rat lung model. The total recovery of the lipophilic drug propranolol was found to be significantly decreased when dextran was used as oncotic agent instead of albumin. Moreover, the measurement of pulmonary disposition of the potent glucocorticoid budesonide after administration to the air-space or the pulmonary circulation of the isolated rat lung was only feasible in the presence of 4.5% albumin in the perfusion medium.

These studies emphasize the use of albumin as oncotic agent in perfusion buffers when the pulmonary disposition of hydrophobic drugs is under investigation. In addition, depending on the experimental design, heparinized animal plasma may be added to the buffer fluid (10–15%). Use of heparinized animal blood as perfusion fluid most closely resembles the *in vivo* state. However, analysis of drugs and interpretation of results are rendered much more difficult by the presence of such a "complex" perfusion medium, thereby negating some of the advantages of the isolated lung technique.

The use of IPL preparations to study the lung disposition of several inhaled therapeutics was pioneered by Byron *et al.* and Ryrfeldt *et al.* in the mid 1980's. The pulmonary absorption and distribution of low molecular and high molecular weight drugs was addressed in several studies. Ryrfeldt *et al.* investigated the pulmonary disposition of the glucocorticoid budesonide in an isolated rat lung after instillation. The drug absorption from the air-space into the perfusate was characterized by two distinct phases: after a rapid initial absorption phase about half of the instilled dose was slowly transferred into the perfusate.

This study points out the high lung affinity of budesonide, no biotransformation of this compound was found in the lung. The high affinity to the lung together with an absence of lung metabolism was shown to be an important factor to explain the clinical benefits seen with budesonide. Kröll *et al.* used the isolated guinea pig lung model for the measurement of the pulmonary fate of two antiasthmatic drugs (xanthines).

After intratracheal instillation of theophylline, the peak concentration in the lung perfusate appeared within a short period of time, and after 10 and 60minutes, ~68 and ~87% of the given dose had been absorbed, respectively. The rapid disappearance of locally administered theophylline may explain the lack of success of inhalation therapy with this therapeutic agent. The pulmonary disposition of the antibiotic levofloxacin was evaluated after systemic application and inhalation in a model of the isolated rat lung. Different experimental conditions including higher or lower respiratory frequency with lower or higher tidal volume were tested. Comparison of systemic and pulmonary administration revealed statistically significant differences between partition coefficients showing much higher values for the latter route. Thus, inhalation compared to systemic administration improves levofloxacin access to the lung tissue.

Only limited information is available regarding the administration, deposition, and absorption of dry powder aerosols to IPL preparations. For this purpose, Ewing *et al.* and Byron *et al.* established an isolated rat lung model and reported the absorption profiles of a variety of test compounds. Using the recently developed DustGun aerosol technology, Ewing *et al.* exposed the IPL model to respirable dry powder aerosols of three drugs at high concentrations. Other interesting techniques for reproducible aerosol application to IPL preparations include the miniaturized nebulization catheter (AeroProbe) and the "forced solution instillation" technique.

Lahnstein *et al.* investigated the pulmonary absorption and distribution of fluorescent dyes in an isolated rabbit lung model. Three structurally diverse probes were administrated intrapulmonary by nebulization of dye solutions. The authors found that the absorption of the model compounds from the air-space into the perfusate was mainly affected by the physicochemical properties (octanol/water partition coefficient) of the employed dyes. While for the hydrophobic dye only a marginal appearance in the perfusate was observed due to accumulation in the lung tissue, a rapid increase in perfusate concentration (with stable plateau concentration) was obtained for both hydrophilic dyes.

Rapid absorption and capacity-limited metabolism of isoproterenol and prodrugs thereof were observed following intrabronchial and aerosol administration of drug to the isolated rabbit lung. The IPL preparation is a valuable model for the analysis of pharmacokinetic profiles of pulmonary administered drugs. Upgrading of the *ex vivo* models to pharmacodynamic investigations has recently become technically feasible. As an example, the pharmacokinetics and vasodilatory effect of nebulized iloprost were investigated in a model of experimental pulmonary hypertension employing the isolated rabbit lung. The nebulization of different amounts of iloprost caused a dose-dependent pulmonary vasodilatation. In addition, a similar dose-dependent appearance of iloprost in the recirculating perfusate was noted.

The effect of polymeric nanoparticles on the microvascular permeability and translocation across the alveolar barrier was tested in isolated rabbit lungs after nanoparticle instillation. The increase in pulmonary microvascular permeability was related to the number of administered nanoparticles. Moreover, positively charged nanoparticles were more effective in the microvascular permeability response than negatively charged nanoparticles. The authors concluded that the surface properties and the total surface area need to be considered to interpret the changes of the microvascular permeability upon nanoparticle challenge.

The applied polymeric nanoparticles were mainly located in the alveolar space and in macrophages after instillation, and no translocation of nanoparticles from the alveoli into the perfusion medium was observed. However, the relevance of these findings for the *in vivo*translocation of inhaled ultrafine particles remains to be established, owing to the fact that polymeric nanoparticles are currently under investigation as potential drug delivery systems to the lung.

Beck-Broichsitter *et al*. compared the pulmonary disposition characteristics of the hydrophilic model drug 5-carboxyfluorescein after aerosolization as solution or entrapped into polymeric nanoparticles in an isolated rabbit lung model. Nanoparticles were of spherical shape with a mean particle size of ~200nm. Nebulization of the nanosuspension using a vibrating mesh nebulizer led to negligible changes of nanoparticle properties. The drug release *in vitro* was fast. Nevertheless, after deposition of equal amounts of 5-carboxyfluorescein in the isolated rabbit lung model, less 5-carboxyfluorescein was detected in the perfusate for loaded nanoparticles (~10ng/ml) when compared to 5-carboxyfluorescein aerosolized from solution (~18ng/ml).

Although IPL studies are conducted with an intact organ close to the physiological state, it is to a large extent unresolved if pharmacokinetic data obtained from IPL preparations are consistent with that measured *in vivo*. Recently, a linear relationship between the drug absorption kinetics of diverse low molecular weight drugs (<700g/mol) in a vertically positioned IPL system and from the lung *in vivo*was demonstrated. Moreover, drugs for which air-to-perfusate absorption kinetics were evaluated *ex vivo* and *in vivo* were also tested in epithelial cell culture models (Caco-2 and 16HBE14o) regarding their transport characteristics. Permeability in intestinal and airway cell culture models were found to be in excellent agreement with the physicochemical properties of the investigated drugs, as well as the rate of absorption measured in the IPL.

The absence of a bronchial circulation in horizontally positioned IPL preparations and therefore a lack of tracheobronchial absorption pathways have been attributed as the likely cause of a substantial difference in the absorption kinetics of low molecular weight drugs between the IPL preparation and *in vivo*.

The absorption of low molecular weight drugs *in vivo* takes place from alveolar, as well as tracheobronchial regions at effective and comparable rates. As a result, the IPL preparation underestimates the absorption of low molecular weight drugs compared to the *in vivo* situation. However, macromolecules show poor to insignificant absorption across the thicker tracheobronchial membranes and thus, the absorption profiles of macromolecules derived from IPL preparations have been reported as statistically indistinguishable from those obtained *in vivo*. Clearly, studies regarding the difference between the vertically and horizontally positioned IPL systems on drug absorption need to be carried out.

In recent studies, the development and performance of a novel nanoparticle-based formulation for pulmonary delivery of salbutamol has been characterized systematically through all steps beginning from the particle preparation process, over the *in vitro* testing of drug release, drug transport in cell culture, pulmonary absorption, and distribution characteristics in an isolated rabbit lung model, to *in vivo*bronchoprotection studies in anaesthetized guinea pigs. Sustained salbutamol release from the drug-loaded nanoparticles was observed for 2.5h *in vitro*. Drug transport experiments conducted with primary cultured human alveolar epithelial cells revealed a delayed transport of salbutamol across the cell monolayer for nanoparticle formulations. In parallel, a sustained salbutamol release profile was observed after aerosol delivery of nanoparticles to the IPL as reflected by a distinct absorption profile and lower salbutamol recovery in the perfusate (~40%) when compared to salbutamol solution (~63%).

Moreover, a prolonged pharmacological effect was observed for 120 min *in vivo* when salbutamol-loaded nanoparticles were administered to guinea pigs. Overall, these results demonstrate good agreement between *in vitro, ex vivo,* and *in vivo* tests, serve as examples for the potential of the IPL to be used to predict drug absorption from the intact animal and, therefore, present a solid basis for future advancement in nanomedicine strategies for pulmonary drug delivery.

Index

A

Abnormalities 90
Abraxane 278
Accelerate 2, 5, 126, 162, 185, 202, 225, 232, 297
Acceptable Methods 83
Accumulation 71, 87, 88, 112, 122, 125, 127, 128, 140, 147, 178, 179, 180, 199, 203, 283, 294
Accusations 12
Aerosolization 278, 283, 286, 295
Aerosolized 296
Afforestation 152, 153
Agglomeration 37, 55, 58, 62, 63, 65, 66, 100, 242, 245, 291, 292
Aggregation 50, 53, 85, 113, 114, 167, 168, 169, 170, 240, 255, 286, 287, 288
Agroforestry 140, 147, 149, 151, 158
Agrosilviculture 147
Agrosilvipasture 147
Alkylammonium 198
Amazingly 20, 23
Amendments 166
Ammonotelic 104
Anaesthetized 297
Anesthetized 290
ANOVA 262
Antimicrobial 46
Assembler 13, 14, 17, 18, 19, 20, 21, 22, 23, 24, 25
Asthmatics 84
Autoregulation 106

B

Bacteria 116, 117, 128, 160, 163, 168, 249, 279
Bactericidal 46
Balcony 20
Bastards 12
Bicarbonate 82, 109, 110
Bioavailability 86, 251, 252, 253, 256, 257
Biodegradable 7, 29, 257, 283, 286, 287, 288
Biomedical 12, 30, 31, 33, 215, 268, 283
Biomonitoring 267
Biopersistence 38, 42, 52, 86, 88, 93, 94, 96, 97

Boiremediation 253
Bradyrhizobia 163
Bronchotracheal 279
Bulk 24, 32, 33, 39, 41, 43, 47, 49, 51, 52, 54, 57, 60, 99, 127, 178, 195, 198, 215, 217, 223, 229
Bulldozing 25
Bumping 25

C

Camouflage 11
Carbon Nanotubes 235
Carbon nanotubes 30, 33, 35, 39, 45, 46, 48, 67, 217, 235
CCN 182, 197, 199
CNC 183, 184, 185, 186, 192
Coagulation 83, 177, 179, 180, 280
Combustion 43, 44, 177, 186, 194
COMET 63
Commercialization 50, 215
Complementary methods 238
CPC 192, 193
Crystallinity 80
Cytotoxicology 264, 265, 266, 270

D

Decomposition 110, 111, 112, 113, 114, 118, 120, 125, 126, 127, 128, 129, 131, 134, 135, 136
Degradative pathways 262
Derivatization 205
Durability 28, 38, 52, 75, 78

E

Earthworms 116, 117, 118, 137, 144, 160, 161, 166, 167, 168
Elasmobranchs 103
Eliminated 4, 91, 104, 134, 137, 227
Embedded 22, 231, 273
Embryonic 74, 268
Emulsifying 100
Encapsulated drugs 277
Endangered 12
Endocytosis 84, 256, 257, 280
Endothelial 62, 64, 65, 70, 73, 88, 90, 284
Endothelium 38, 70, 79, 96
Endpoints 62, 63, 64, 65, 67, 70, 71, 72, 73, 74, 77, 81, 83, 86, 94, 97
Epidemiological 43
Epidermal 45, 66, 67, 68, 259
Epiphaniometer 192
Epithelial cells 45, 62, 64, 65, 78, 103, 256, 260, 278, 280, 289, 297
Epithelium 62, 63, 69, 82, 87, 103, 108, 279
Equivalent 125, 170, 177, 178, 189, 201, 269
ERXPS 242
Escherichia coli 261
Eyebrow 142

F

Fabricated 32
FASET 205
FDA 31
Fertilization 142, 156
Fibrinoylsis 280
Fibroblasts 62, 64, 78
Fluorescence 46, 68, 202, 258, 260, 261, 262, 273, 274
Foreseeable 264
FTIR 186, 236
Fuelwood 150, 152, 153
Functionalization 236, 237
Fungi 113, 116, 117, 128, 144, 160, 170, 249, 259

G

Genotoxicity 63, 75, 79
GFR 106, 107
Glomerular 105, 106, 107, 109
Glutathione 63, 272

H

Hepatocyte 47, 71, 72, 265
Homogeneous 40, 180, 185, 189, 273, 285, 290, 292
HRTEM 194
Hyperhumid 126

I

Immobilization 128, 157, 274
Immobilized 239
Immortalized 69, 72, 74, 259
Immunological 65, 281
Immunopathology 71, 79
Inaccuracies 19
Inflammation 38, 42, 44, 45, 63, 71, 75, 79, 81, 94, 97
Inmobilized 166
IPPSF 68, 79

J

JGA 106
JWG 217

K

Keratinocytes 45, 66

L

Liberating 136

M

Machinery 1, 2, 17, 19, 20, 22, 23, 164
Maintaining 103, 109, 117, 129, 137, 159
Mammalian 44, 86, 109, 263
Marshmallow 99
Microemulsion Method 273
Microorganisms 111, 120, 124, 139, 145, 148, 163, 251, 252
Microwatershed 142
Miniaturization 32
MMAD 285, 287
Modalitie 278
MWCNT 45, 235, 237
Mycorrhizae 153

N

NCL 30, 31
Nebulization 286, 291, 294, 295
Neutralization 188
NIST 31
Nucleation 177, 180, 182, 183, 184, 188, 189, 198, 199, 200

O

Occlusion 56, 92
Oligodendrocytes 73
Oxidation 71, 81, 100, 113, 128, 198, 199, 202, 203, 204, 224, 225, 227, 228, 233, 236, 237, 242

P

Peptization 99
Permeability 35, 69, 74, 81, 102, 106, 107, 108, 280, 289, 295, 296
Phagocytosis 63, 64, 66, 87, 256, 257, 258, 259, 261, 262

Pharmacokinetic 277, 282, 289, 291, 292, 295, 296
Pharmacological 276, 277, 297
Phosphatidylglycerols 279
Phosphatidylinositols 279
Phospholipids 279
Physicochemical 31, 37, 40, 42, 43, 48, 49, 50, 51, 54, 57, 59, 60, 85, 95, 189, 249, 252, 253, 254
PLA 283
Plundered 9
PNAPs 287, 288
Poisonous 25
Polymerization Method 273
Possibilities 5, 19, 33
Precipitators 101
Prediction 197, 263, 265, 278, 289
Preexisting 189, 197
Prioritization 57, 76
Programmable 14
PROPHET 202, 204, 205, 207
Pulmonary 38, 43, 44, 65, 80, 81, 82, 83, 84, 90, 91, 92, 93, 94, 97, 275, 276, 277, 278, 279, 280
Pulverization 137, 162

R

Ramifications 31, 46
Reabsorption 105, 106, 107
Realization 264, 265
Recirculation 290
Reduction 67, 69, 99, 100, 122, 126, 130, 131, 136, 137, 138, 143, 170, 173, 174, 181, 182, 210
Rehabilitation 145, 152
Reproductive 38, 90, 94, 97
Rhizosphere 248, 249, 250, 251, 252, 253, 254
Ricocheting 3, 26

S

Salbutamol 297
Separation Method 193
Silvipasture 147
SIMS 186, 219, 231, 232, 233, 239, 245, 246, 247
Smallholder 147, 164
SNOM 191
Soilbinding 169
SOM 250
Sophisticated 1, 29, 32, 48, 288, 292
Spacecraft 1, 12, 26, 34
Spectroscopies 219, 222, 224, 232, 236
SPM Methods 219, 234, 247
Spontaneous 99, 133, 259, 287, 288
Stabilizers 283, 287
Standardized 47, 59, 265
Styrofoam 99
Susceptibility 84, 85, 148
SWCNT 45, 235, 236, 237

T

TCNQ 237
TDCIMS 188, 199, 200
Tdcims 201
Terribly 21, 23
Therapeutic 30, 256, 275, 276, 282, 283, 285, 286, 287, 288, 289, 291, 292, 293, 294
Thoracotomy 290
Toxicological 39, 41, 43, 44, 45, 47, 66, 191, 195, 249, 252, 254, 270, 277

Tracheobronchial 296
Transducers 290
Transfection assays 262
Translocation 38, 52, 63, 69, 70, 74, 77, 83, 85, 86, 87, 88, 93, 94, 96, 97, 280, 295
Transplantations 71
Trapezoidal 142
Trihalomethane 174
Tumbling 25
Tweezers 15

V

Vasodilatation 295
Vitroexperiments 44

W

Wandering 25
WPMN 216